SET ADRIFT:
FISHING FAMILIES

Set against the backdrop of the fisheries crisis of the 1990s, *Set Adrift* examines how coastal and deep-sea fishermen's wives in rural Nova Scotia have adapted to the extraordinary pressures put on their households by the reorganization of the fishing industry. Using in-depth interviews conducted with the wives of deep-sea and coastal fishermen, members of fishermen's wives' support groups, and fish company managers, Marian Binkley explores the role of social origins and family traditions, family and social networks, and the availability of employment opportunities and social services on fishing households.

Comparing and contrasting the households of deep-sea and coastal fishers, Binkley illustrates the daily dependence of husbands upon their wives' labour and ability to adapt to often difficult and precarious living conditions. Maintaining that women make the fishing industry sustainable with their unpaid household labour, Binkley argues that the failure of Canadian government officials and policy makers to recognize the centrality of women's labour to the industry has resulted in fishers' wives bearing the brunt of the large economic and social costs generated by the current fisheries crisis. Ultimately, she contends, any analysis of production for exchange must recognize the essential contribution that household domestic labour makes to the sustainability of economic activity.

MARIAN BINKLEY is Professor of Anthropology and Dean of the Faculty of Arts and Social Sciences at Dalhousie University. She is also the author of *Risks, Rewards and Dangers in the Nova Scotia Offshore Fishery* and *Voices from Off Shore: Narratives of Risk and Danger in the Nova Scotian Deep-Sea Fishery*.

MARIAN BINKLEY

Set Adrift:
Fishing Families

UNIVERSITY OF TORONTO PRESS
Toronto Buffalo London

© University of Toronto Press Incorporated 2002
Toronto Buffalo London
Printed in Canada

ISBN 0-8020-4812-9 (cloth)
ISBN 0-8020-8351-X (paper)

∞

Printed on acid-free paper

National Library of Canada Cataloguing in Publication

Binkley, Marian, 1950–
  Set adrift : fishing families / Marian Binkley.

  Includes bibliographical references and index.
  ISBN 0-8020-4812-9 (bound).   ISBN 0-8020-8351-X (pbk.)

  1. Fisher's spouses – Nova Scotia – Social conditions.   2. Fishers'
  spouses – Nova Scotia – Economic conditions.   3. Fishers –
  Nova Scotia – Family relationships.   I. Title.

  HD8039.F652C3 2002   305.48'9653'088639209716   C2002-903137-0

This book has been published with the help of a grant from the Humanities and Social Sciences Federation of Canada, using funds provided by the Social Sciences and Humanities Research Council of Canada.

University of Toronto Press acknowledges the financial assistance to its publishing program of the Canada Council for the Arts and the Ontario Arts Council.

University of Toronto Press acknowledges the financial support for its publishing activities of the Government of Canada through the Book Publishing Industry Development Program (BPIDP).

For the women of my family:

Margreta Mary-Anne de Bursey Binkley Andrews
Ina Henrietta de Bursey
Sarah Maude Hurst Fitzpatrick
Sarah Elizabeth Fitzpatrick Binkley
Mary Elizabeth Foley Crowley
Amy Elizabeth Crowley
Hannah Elizabeth Crowley Wallworth

Contents

List of Tables and Figures ix
Acknowledgments xiii

1 Introduction 3
   Another Bust in the Fisheries 5
   Living with a Fisherman 12
   The Study 15

2 Living the Dream 22
   Dreaming of One's Own Boat 23
   Achieving the Dream 24
   Helping Out 28
   The Dream in Peril 36

3 Two Separate Worlds 43
   His World 44
   Her World 47
   Intersecting Worlds 52
   Downsizing, Work Reduction, Retirement 57

4 Running the Household 65
   Dividing Up Household Tasks 65
   Looking After Children 72
   Helping Out around the House 80
   Getting Him More Involved 86

viii   Contents

5 Family, Friends, Acquaintances   91
  I Can Always Depend on My Family, But at What Cost?   93
  My Friends Help Out   101
  I Can Call on My Neighbours in a Pinch   105

6 Just Having Fun   112
  Going Out and Staying In   117
  Having a Few Drinks   121
  Drinking, Stress, and the Job   131

7 Going to Work   134
  The Life Cycle and Employment Histories   141
  Staying Home   142
  Who's Working Now?   148
  I Want My Own Money   153

8 Our Money, Your Money, My Money   158
  Why Plan?   163
  Who Has Access to These Resources?   167
  How Are Decisions Made?   171
  How Do These Households Work?   178

9 Conclusions   186
  What of the Future?   190

Appendix   195
  Table 1 Demographic Profiles of Deep-Sea Fishermen's Wives at Time of Interview   196
  Table 2 Demographic Profiles of Coastal Fishermen's Wives at Time of Interview   197
Notes   199
References   205
Index   215

Tables and Figures

Tables

1.1 Demographic Profiles of Samples (in percentages)   19
2.1 Reported Involvement of Wives in Their Husbands' Fishing Enterprises: Comparison of Coastal Sample and Findings of Theissen, Davis, and Jentoft (1992) (in percentages)   30
2.2 Reported Involvement of Wives in Their Husbands' Fishing Enterprises: Comparison among Coastal Wives by Husband's Job Status (in percentages)   31
2.3 Reported Involvement of Wives in Their Husbands' Fishing Enterprises: Comparison among Coastal Captains' Wives by Wife's Employment Status (in percentages)   34
4.1 Frequency of Domestic Chores Done by Women for Coastal Sample (in percentages)   67
4.2 Frequency of Domestic Chores Done by Women for Deep-Sea Sample (in percentages)   67
4.3 Frequency of Use by Women of Community Services for Coastal Sample (in percentages)   70
4.4 Frequency of Use by Women of Community Services for Deep-Sea Sample (in percentages)   70
4.5 Distribution of Number of Children per Family (in percentages)   73
4.6 Age Profile of Children by Type of Fishing Enterprise (by frequency and percentage)   73
4.7 Age Profile of Children by Type of Fishing Enterprise Where Mother Was Employed (by frequency and percentage)   74

x   Tables and Figures

4.8   Primary Childcare Giver for Pre-school-age Children (in percentages)   75
4.9   Type of Minders for School-age Children by Sample (in percentages)   77
4.10 Husbands' Interaction with Pre-school-age Children (in percentages)   78
4.11 Comparison of Husbands' and Wives' Interaction with School-age Children (in percentages)   79
5.1   Place of Residence at Marriage for Coastal and Deep-Sea Samples (in percentages)   98
5.2   Distance from Kin for Coastal and Deep-Sea Samples (in kilometres)   99
5.3   Indicators of Wives' Belonging to Community for Coastal and Deep-Sea Samples (in percentages)   105
6.1   Frequency of Wives' Involvement with Various Forms of Entertainment for Coastal Sample When Husband Is at Home and Away (in percentages)   113
6.2   Frequency of Wives' Involvement with Various Forms of Entertainment for Deep-Sea Sample When Husband Is at Home and Away (in percentages)   114
6.3   Frequency of Wives' Involvement with Community Organizations for Coastal Sample When Husband Is at Home and Away (in percentages)   115
6.4   Frequency of Wives' Involvement with Community Organizations for Deep-Sea Sample When Husband Is at Home and Away (in percentages)   116
6.5   Comparison of Activities with Other Fishing Couples for Study Samples (in percentages)   119
6.6   Comparison of Women's Activities with Other Fishermen's Wives and Workmates for the Study Samples (in percentages)   121
6.7   Frequency of Women's Drinking While Participating in Various Social Activities in the Year Preceding the Survey (in percentages)   122
6.8   Comparison of Coastal and Deep-Sea Fishermen's Drinking Patterns Inside and Outside the Home (in percentages)   124
6.9   Frequency of Fishermen and Their Wives Engaging in Binge Drinking (in percentages)   126
6.10 Comparison of Consequences of Other People's Drinking on Women Based on Data from Both This Study and the National Alcohol and Other Drugs Survey (in percentages)   128

Tables and Figures   xi

6.11 Reported Frequency of the Type of Help Given Fishermen's Wives Related to Husbands' Drinking (in percentages)   130
7.1   Characteristics of Women's Employment by Sample (in percentages)   135
7.2   Frequency of Reported Income Distribution of Working Wives of Coastal Fishermen Broken Down by Employment Status (in percentages)   136
7.3   Frequency of Reported Income Distribution of Working Wives of Deep-Sea Fishermen Broken Down by Employment Status (in percentages)   136
7.4   Frequency of Reported Income Distribution of Working Wives Broken Down by Educational Status for Deep-Sea and Coastal Fishermen's Wives (in percentages)   137
7.5   Event Marking Homemaker Status for Deep-Sea and Coastal Fishermen's Wives (in percentages)   143
7.6   Husband's Support for Wife's Work Status Reported by Deep-Sea and Coastal Fishermen's Wives (in percentages)   145
7.7   Most Important Quality of Their Job for Deep-Sea and Coastal Fishermen's Wives by Sample Ranked According to Coastal Sample (in percentages)   149
7.8   Compilation of Three Most Important Qualities of Their Job for Deep-Sea and Coastal Fishermen's Wives by Sample Ranked According to Coastal Sample (in percentages)   150
8.1   Frequency of Reported Income Distributions of Coastal Fishermen Broken Down by Job Status (in percentages)   160
8.2   Frequency of Reported Income Distributions of Deep-Sea Fishermen Broken Down by Job Status (in percentages)   160
8.3   Sources of Husbands' Reported Income for Fishing-Dependent Households Broken Down By Deep-Sea and Coastal Fishermen (in percentages)   161
8.4   Distributions of Personal Earnings Reported by Wives of Deep-Sea and Coastal Fishermen (in percentages)   162
8.5   Household Income Distributions Reported by Wives, Broken Down by Household Type (in percentages)   162
8.6   Deep-Sea and Coastal Fishing-Dependent Households' Use of Different Types of Bank Accounts, Credit Cards, and Charge Accounts (in percentages)   168
8.7   Use of Segregated versus Pooled Resources by Deep-Sea and Coastal Fishing-Dependent Households (in percentages)   170
8.8   Management of Household Accounts by Wife and/or Husband in

Tables and Figures

Deep-Sea and Coastal Fishing-Dependent Households (in percentages)   172
8.9 Spouses' Decisions on Daily and Major Purchases (in percentages)   174
8.10 Wife's and/or Husband's Control of Household Accounts (in percentages)   176

Figures

1.1 Map of Atlantic Canada   7
1.2 Map of Nova Scotia Showing the Study Area   8

Acknowledgments

Set Adrift grew out of my earlier research with Nova Scotian deep-sea fishermen and their households. As part of that study, I interviewed twenty-five wives of deep-sea fishermen, and those interviews formed the basis of an article I wrote with Victor Thiessen (1988). When I presented an early version of that paper to a women's studies seminar at Dalhousie University, two women came up afterwards and spoke about the similarities and differences between their lives as wives of coastal fishermen and those of the wives I had described. That informal chat sparked this project.

The Social Sciences and Humanities Research Council of Canada (410-92-0518) and the Dalhousie University Research and Development Fund supported the research for *Set Adrift*. The research team's composition changed dramatically during the course of the study. In Phase 1, Meredith Ralston, Connie Wawruck-Hemmett, Maggie Macdonald, Valerie Noftle, and Kim Short assisted me in compiling a review of the literature; conducting informal, in-depth interviews with key informants; identifying possible participants for the study; and developing statistical, demographic, and socio-economic profiles for the study area. They also helped in designing the interview schedules and sampling plans for subsequent phases of the study. Phases 2 and 3 consisted of 300 interviews with fishermen's wives and 50 taped follow-up interviews. Marie Patural was hired as project manager to oversee these phases of the study. Ariella Phalke, Adriana McCrea, Lynn Langille, and Marie Patural assisted with the interviews. Michael Peckham and Wade Deisman coded and entered the data from the surveys. Mary-Ann Finlayson transcribed the 50 follow-up interviews. Phase 4, analysis and writing up the study, was my responsibility alone.

A number of people from government agencies – the Department of Fisheries and Oceans, and the Nova Scotia Ministries of Fisheries and Labour – contributed to this project. I would particularly like to thank Leo Brander, Janice Raymond, and Brian Thompson. Throughout, many colleagues gave me encouragement and advice: Richard Apostle, Dona Davis, Siri Gerrard, Jennifer Jarman, Meg Luxton, Barbara Neis, and Victor Thiessen. During the writing stage, the Huntington Library, Art Collections, and Gardens in San Marino, California, gave me space to work on the manuscript.

This book would not have been completed without the good-humoured prodding of my husband, Jack Crowley. His enthusiasm for the project, and his willingness to read and discuss earlier drafts of the manuscript, made the task of writing it a more enjoyable endeavour. A task shared always lightens the load, and so it was in this case.

Finally, this project could not have been undertaken, or the book realized, without the support of the people who were its focus. To them I extend my warmest thanks, and I wish them the best in their struggle to make a living from the sea and to care for their families. I particularly want to single out the women of Fishnet, an Atlantic Canadian organization whose efforts have raised the public's understanding of the plight of fishing families throughout Atlantic Canada.

**SET ADRIFT:
FISHING FAMILIES**

CHAPTER ONE

Introduction

Living with a fisherman, you have to realize that you have to be very, very independent, have your own life, and don't rely on him to make your life. You have to be happy with yourself and accept him for what he is in his life because they are a special breed of people (laughter), they really are. (Oriel)[1]

Set Adrift began in 1992 as an empirical study of the experiences of fishermen's wives, but the research was soon set against the backdrop of a fishery in crisis in the 1990s.[2] This book compares the adaptations of two groups – the wives of coastal fishermen and the wives of deep-sea fishermen – in response to the extraordinary pressures put on their households by the changing nature of their husbands' work. It also examines the ways Nova Scotian fishing-dependent households differed, and focuses on women's contributions to these households as they attempted to cope with the economic and social restructuring of their communities in response to the declining fish stocks.

Until the 1990s, social sciences literature marginalized the discussion of gender relations and women's roles in the fishing industry and household fishing enterprises. Many researchers viewed the contributions of women to these endeavours as secondary to an understanding of men's work in the fishing industry or of the fishing industry itself. Only Charlene Allison and co-authors (1989) explored the role of women as fishers. The other notable exception was the edited monograph by Jane Nadel-Klien and Dona Davis (1988), *To Work and to Weep: Women in Fishing Economies*. And, as Davis and Nadel-Klein (1992) have pointed out, 'much of general maritime studies was highly androcentric. Those scholars who did mention women frequently relegated them to a passing comment, paragraph, or discrete section on household and/or family. Accounts focusing on women as major actors in fishing economies were relatively rare' (135).

Research on women and the fisheries grew substantially throughout the 1990s.[3] The resulting literature has focused on women's contributions to their household's fishing enterprise through unpaid work; on the relationship between women's paid and unpaid work; and on the recognition of these contributions, or lack thereof, by government, community, and individuals. It discusses as well the gender division of labour in paid work within all sectors of the fishing industry – harvesting, processing, and marketing – and the gender inequities that existed in wages, working conditions, and access to social benefits and income support. It also explores household dynamics, such as the division of labour and access to resources within the household, as well as the relationship between gender and the effects of globalization and restructuring on the fishing industry against the backdrop of declining fish stocks worldwide (Davis and Gerrard 2000). Despite all of this fine work, gender issues remain on the margins (Harrison 1995). This literature is read by feminists and progressive thinkers in fishery research, but it remains peripheral to the mainstream works in the field (e.g., Apostle *et al.* 1998; Arnason and Felt 1995; Newell and Ommer 1999).

Set Adrift centres on gender issues. It focuses on women's work and their contributions to the fishery in the 1990s. It shows that women's participation in the fishing endeavour and in wage labour contributed to the household income and directly supported the fishing endeavours. It explains how women's labour in the fishery varied according to their husbands' occupation as coastal or deep-sea fishermen, and their husbands' status aboard ship. Moreover, it demonstrates that women's social and emotional support was indispensable to maintaining the fishery. It argues that without women's labour these fishing endeavours could not have functioned, that women's labour was crucial, not tangential, to the functioning of the fishing industry at and beyond the subsistence level.

As the study evolved, an important change took place. The Atlantic Canadian cod fishery experienced a catastrophic decline, which by 1993 resulted in a near cessation of all cod fishing in Atlantic Canada. In those areas exempt from this moratorium – the waters off the shores of southern Nova Scotia, from Halifax to Yarmouth, Georges Bank, and the Bay of Fundy – the federal government severely curtailed the cod fishery and other groundfisheries (e.g., pollock, halibut, and haddock), but did not close them. The research area for this book fell within this exempt region. The households represented in the study experienced substantial disruption to their way of life, but were not as severely affected as those fishing-dependent households where the moratorium remained in place.

Thus what began as a descriptive study of a social phenomenon defined by an occupational group (fishermen's wives) then became a documentation of social change of households in crisis, an experience not unique to the fisheries. Although the research problem still focused on fishermen's wives and fishing-dependent households, its definition took on an additional dimension. The study became a story about the changes wrought by the fisheries crisis, the subsequent changes made in household adaptations, and the negotiation of the roles of husbands and wives in light of these changes. That is, it became a further example of how the restructuring of a primary industry affects the people who depend on that industry for their livelihood, and in turn how these changes can impact families and communities.

The fishery sector traditionally employed, directly or indirectly, the largest number of workers in Nova Scotia, a region already burdened by high unemployment. Fishing, combined with employment in onshore processing plants and in related enterprises, formed the economic and social backbone of the small communities nestled along the province's indented Atlantic coastline. The rise and fall of the fortunes of the fishery dictated the economic prosperity or decline of the communities dependent on it.

Another Bust in the Fisheries[4]

Although the history of the Atlantic Canadian fishery is a tale of boom and bust, the impact of the crisis of the 1990s rivaled the Great Depression of the 1930s and was part of a global problem. The world fisheries were in crisis: marine resources were in global decline (McGoodwin 1990; UN–FAO 1998). Around the world, maritime communities, previously dependent on marine resources, were dramatically downsizing their fisheries, and many were turning to other industries, most notably tourism. Like all communities undergoing massive restructuring, maritime communities were altered culturally (Appadurai 1990), economically (Watts 1992), and politically (Harvey 1989). The processes of globalization and restructuring reduced the subsistence production of small-scale fishers as large multinational companies controlled access to more and more of the marine resources. As more fishers and their families were deprived of their access to these resources they increasingly depended on wage labour and government welfare benefits (Nash 1994); yet these same governments decreased social benefits. The few small-scale fishers who managed to retain their access to marine resources were economically squeezed and became dependent on the large multinationals for the sale and processing of their products in the global market. Fishing-dependent communities also paid the price for this economic remedy through higher unemployment rates, increased emigration, higher costs of living, and increased de-skilling of labour (Apostle et al. 1998; Watts 1992).

Throughout the North Atlantic fish stocks declined dramatically. The latest crisis in the region began in the 1980s and peaked in the early 1990s (Arnason and Felt 1995). Although scientists speculated on a variety of ecological and environmental reasons for the decline, the chief cause was overfishing – too many men chasing too few fish (Finlayson 1994). Throughout the region, governments responded to the crisis by putting in place management policies which increased the economic efficiency of the fishing industry while promoting sustainability of the marine resources.

When I began an earlier study relating to the industrial fishery in 1986, 54,084 fishermen and 22,133 plant workers worked in Atlantic Canada (Canada, DFO 1991: 111–13). In 1987, the Atlantic Canada fishery accounted for 3 per cent of the world's total catch, 80 per cent of Canada's total catch, and two-thirds of Canada's total fish exports (Northwest Atlantic Fisheries Organization 1989; United Nations Food and Agricultural Organization 1987). Between February 1992 and August 1993 the Canadian federal government gradually halted harvesting of northern cod off Newfoundland's shores, in the Gulf of St Lawrence, including Sydney Bight, and in waters off Nova Scotia north of Halifax, virtually prohibiting all cod fishing off the Atlantic shore north of Halifax except for fisheries off the Labrador coast (see Figure 1.1). In the remaining areas fishing continued, but with greatly reduced quotas for groundfish, including cod and pollock. The moratorium led to a drastic restructuring of both the fishing industry and the communities that supported it and were supported by it. As of 1993, the northern cod closures had eliminated over 40,000 fishery jobs (Kelly 1993: 14–15).[5] Susan Williams (1996: 1) estimated that in Atlantic Canada 50,000 people working in the fishing industry and another 47,000 people working in fishery-dependent sectors saw their employment modified by the fisheries crisis.

The Canadian federal government also made substantial changes to

Introduction 7

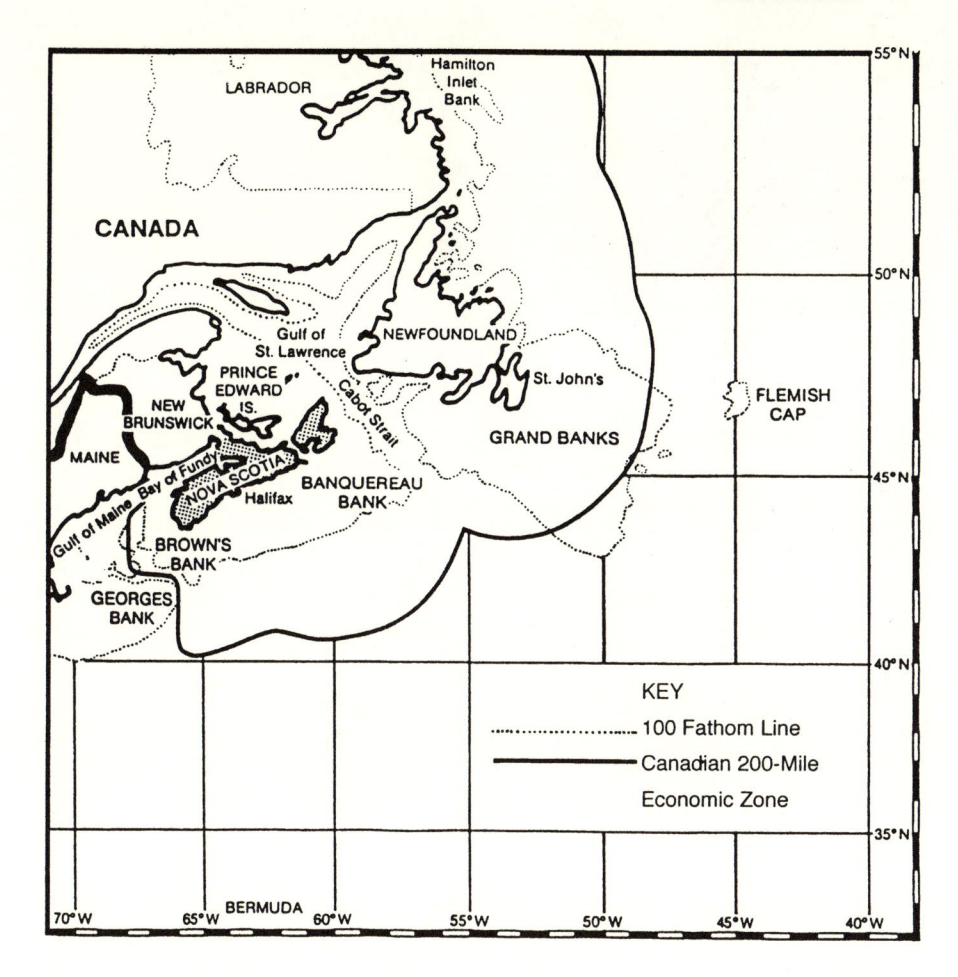

Figure 1.1 Map of Atlantic Canada

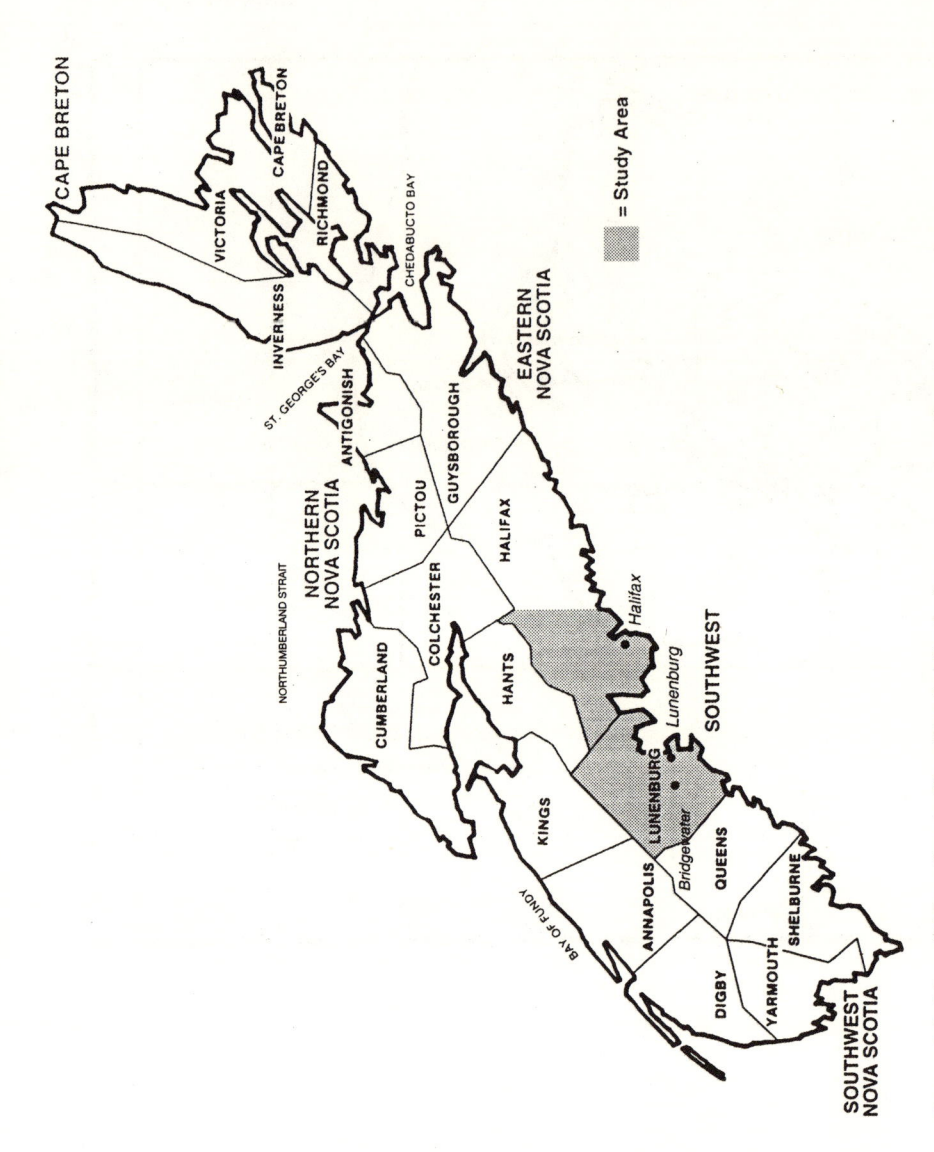

Figure 1.2 Map of Nova Scotia Showing the Study Area

the way the fishery was managed. By withdrawing from its responsibilities in resource management and maintenance of infrastructure within the fishing industry, the federal government reduced its spending in the fishery sector. For example, it increased the costs of fishing licences, passed on some of the costs of research and fishing surveillance to fishers, and gave communities the responsibility for the maintenance of wharves and breakwaters. The stated aim of these and other policy changes was to create a more economically efficient fishery based on reduced but sustainable fish stocks with fewer workers labouring in a more industrialized setting.

At the same time, provincial and federal government reforms of social programs hit fishing-dependent communities hard. Changes in employment insurance and retraining programs, and reduced funding for health, education, and other social benefits met with strong resistance in fishing communities where alternative employment opportunities were few and the ability to transfer current skills to other employment was minimal.

The Canadian government offered some relief in 1995 through a social benefits package called The Atlantic Groundfish Strategy (TAGS). It ended, however, in the summer of 1998, forcing fishing communities to redefine and restructure to meet the new economic reality of both diminished primary resources and changing national and international policies.

People who were laid off or had been forced out of the fishery altogether had to find other sources of income. Some tried to find other employment – not easy for fishers or plant workers with relatively little formal education in a region with chronically high unemployment. Some relied on employment insurance (EI) or social assistance programs, which had buttressed these fishing-dependent communities in the past. However, both the federal and provincial governments decreased support for such programs so they were harder to get and offered fewer resources. Some families were forced to relocate to find employment. The 'Fishermen's Package' (TAGS) offered some support but came with many strings attached. Its retraining programs did not appear to take into account either the saturated regional market or the appropriateness of such programs for people whose formal education had stopped before high school graduation. As it required participants to give up their fishing licences, many fishers were reluctant to sign on. Most of these people planned to return to the fishery once the 'stocks bounce back,' and remained unwilling to give up these hopes. However, if the fishery did rebound no one expected it to support the large numbers of crews and plant workers who had previously exploited these resources. With the conclusion of the TAGS program, the future for fishers, their households, and their communities appeared bleak.

Discussions of the fisheries crisis revolved around the concerns of managing the fish stocks, methods for harvesting and marketing the resources, and plans for the economic development of areas hardest hit by the crisis. Discussion, in consultation with government, marine scientists, economists, and fishers, focused on co-management, quota systems and other marine resources management practices (e.g., Copes 1986; McCay et al. 1995), and on local knowledge systems and their applicability to marine resources management systems (e.g., McCay 1995; Neis et al. 1996). This situation was not unique. Global response to the world fisheries crisis has been similar, including rapid development of aquaculture, technology transfer, importation of scientific management regimes, development of the tourist industry, and the liberalization of trade and of direct foreign investment (Shrybman 1999; Stonich et al. 1997; Wilks 1995). The roles and needs of women in the fishery barely figure in these strategies.

In the case of the North Atlantic Fisheries Crisis, a few researchers (Binkley 1995a, 1996; Christiansen-Ruffman 1995; Davis 1995; MacDonald 1994; Neis 1996; Neis and Williams 1997; Williams 1996) addressed the unique situation of women involved in the crisis. Women's work – paid and unpaid – was drastically affected. For example, of the 35,000 fishers and plant workers in Newfoundland and Labrador who lost their jobs, about 12,000 were women. But Neis (1996: 36) argued that the 'crisis also affected women doing unpaid work in their husbands' fishing enterprises, such as bookkeeping and supplying and cooking for crews. Other women lost their work in childcare and retail sectors in fishery-dependent communities. In addition, out-migration and government cutbacks reduced the number of women employed in education, health, and social services.' These women had a vested interest in their communities, but their cries were not heard in the cacophony of voices shouting for help. Programs set up by the government, such as TAGS, excluded all but female fish-plant workers and a few well-documented full-time female fishers. Moreover, these government-sponsored programs offered training in traditional low-wage jobs for women. There was no acceptance of the contribution women made in the unpaid-work sector of fishing household enterprises, or the loss of employment through secondary economic spinoffs.[6]

Introduction   11

The fisheries crisis in Atlantic Canada went beyond an environmental crisis in resources management to a crisis for coastal communities and their way of life. But Canadian women were unable to affect the debates about fishery policy. In contrast, women of North Norway were able to play a key role in turning the fishery policy debate away from a narrowly defined economic crisis and towards one concerning social and community values. Norwegian women were instrumental in making the links between their ecology, households, employment, markets, and communities a part of the public debates about the crisis (Gerrard 1995). During the crisis of the 1990s, Canadian women failed to make those links part of the public policy debate. As a result, their vital role in the fisheries continued to be invisible and there were few initiatives designed to compensate them.

Set Adrift describes the adjustments fishing-dependent households made to accommodate the economic restructuring of the North Atlantic fisheries, a response to the depletion of the fish stocks. It contributes to the expanding literature, which explored the gendered consequences of recent changes in the fisheries due to globalization and restructuring.[7] It also identifies the types of problems faced by fishing-dependent households, and explores the changes needed to help them meet these challenges.

Set Adrift argues that the fisheries depended on wives' unpaid household labour, and shows how variations in the organization of two distinct types of fisheries – coastal and deep-sea – led to different patterns of household dependency on these women's labour. Government and society failed to recognize the centrality of these women's work to the success of the fishing industry. Women involved in the fishery as shore crew or managers were unable to receive government compensation (e.g., TAGS, employment insurance, retraining packages) for their economic losses related to the fisheries crisis. As the crisis deepened, women's labour, either through more involvement in the fishing industry or through paid employment outside the fishing industry, became even more critical to the economic well-being of fishing-dependent households. Moreover, decreased government spending on health care and education resulted in reduced services, such as closures of hospitals, reduction in medical services, and limited educational programs in their communities. Women, the main caregivers in fishing families and more broadly in their communities, were heavily burdened by these changes. As access to services was reduced, these women tried to compensate by providing alternative services. They also bore, disproportionately, the emotional burden of helping their families cope with the loss of their livelihood. Thus these women paid most of the economic and social costs generated by the fisheries crisis. With these increased burdens and the reduction of governments' social support, both the women and their households felt they had indeed been set adrift.

Living with a Fisherman

How are these struggles played out in everyday life in fishing-dependent households? Let's look at a 'typical day' for two fictitious households: that of Mary and her husband, Bert, a deep-sea fisherman, and that of Jane and her husband, Sam, a coastal fisherman. Both live in the same small community of Captain's Cove. These vignettes illustrate the differences between deep-sea and coastal fishing-dependent households and introduce the various ways households organized their daily routines.[8]

At four o'clock in the morning the alarm goes off and another day begins for the coastal fishing household of Jane and Sam. Jane stumbles out of bed to let the dog out and make Sam's breakfast before he goes to sea. Meanwhile Sam gets up, has a shower, and dresses for work. At the breakfast table, Sam outlines the chores he wants Jane to take care of while he is away. While they talk, Jane makes and packs his lunch. Jerry, Sam's crewman, arrives; after having a cup of coffee Jerry and Sam leave for the wharf. Now Jane has her first cup of coffee and takes a short rest before the day begins in earnest. This is her favourite part of the day, when the house is quiet and the dog is curled up at her feet. It lasts until about a quarter to seven.

Just about then, in Mary and Bert's household, Mary wakes up, has her shower, and gets ready for work. Bert's boat should dock tomorrow, but since his schedule fluctuates, Mary phones the company's hotline. The pre-recorded message lists the status – in harbour, set sail, at sea, or heading for home – of each of the company's boats. The message also gives the scheduled docking date, or if its docking is to be within forty-eight hours, the proposed time. According to the hotline Bert's boat will dock tomorrow morning at seven, so Mary has a day to get ready for his arrival. She wakes up her two teenage boys, Roy and Jim. They shower and get ready for school while Mary makes their breakfast. As the boys eat, she prepares their lunches, reminds them that their father will be home tomorrow, and that they will be celebrating Roy's birthday tomorrow night. While Bert was at sea the older boy, Roy, celebrated his seventeenth birthday with his friends over the weekend, but the family will have a separate celebration when Bert comes home. Mary teaches in the boys' high school, so they all leave for school around eight-fifteen.

Meanwhile, after her shower, Jane wakes her three children – Joe, Kathy and Ann – gets their breakfast, makes their packed lunches, and readies them for school. There always seems to be some emergency – a permission slip that needs to be found and signed, gym shorts that must be mended, homework that must be hastily completed at the breakfast table. Once Jane gets the children on the school bus, she returns home for her second cup of coffee and plans the rest of her day – picking up the bait and the repaired engine part and dropping them off at the wharf, as well as the usual cleaning, laundry, cooking, and grocery shopping. After school Kathy and Ann have music lessons and it is Jane's turn to take Joe and his two friends to swimming classes. Although the GST (goods and services tax) report falls due at the end of the month, Jane decides to put it off until the weekend.

Mary returns home from school about four-thirty. One of the other teachers has dropped her off because Roy has the car. He is doing a number of errands, including picking up the dry cleaning and getting the birthday cake and food for tomorrow. Mary has come home to clean the house and get everything ready for Bert's return. After making supper for herself and the boys, she works all evening getting the last of her marking done so she will be free to be with Bert. He is only home for two or three days, and during that time she wants to have everything just the way he likes it. She will cook his favourite meals, go out to the movies with him, invite his friends over for dinner, perhaps join him for a family occasion. When the boys were younger she did substitute teaching so she could spend the whole shore leave with Bert. But now jobs are scarce, and they need to save money to pay for the boys' university tuition and other expenses, so she teaches full time and can only spend the evenings with her husband.

Meanwhile, Jane's afternoon and evening are just as hectic. The girls' music lessons run late, so she is delayed picking up the boys at the pool. Instead of making the children's supper she takes them to a fast-food restaurant. Once home, the girls do their homework. She gives Joe his bath and gets him ready for bed. Jane then prepares supper for herself and her husband, while the girls watch their one TV show before their baths. As they head for bed, Sam arrives home weary. After tucking the girls into bed and saying goodnight to them (Joe is sound asleep), Sam has his shower. Jane begins washing his fishing clothes for the morning. Over supper, Sam and Jane talk about their respective days. Jane explains that the garage could not fix the part for the boat until tomorrow, so she will have to pick it up then. Sam talks about his problems with the boat's motor, and fears it will have to be overhauled before the end of the lobster season. As Sam heads off to bed, Jane does her mending. She goes to bed herself about ten-thirty or eleven.

In many ways these two households appear very similar to most other Canadian households. In each case, the wife's unpaid domestic labour kept the household functioning. Like the majority of Canadian women, Mary had a job and her wage labour contributed to the household's income. Yet these households were very distinctive because of the way in which the demands of fishing imposed particular constraints on household schedules and divisions of labour.

Bert's job as a deep-sea fisherman compromised Mary's daily schedule for days at a time. When Bert's work took him away from home for substantial periods of time, Mary became fully responsible for the daily running of their home and the care of their children. She reorganized her activities to accommodate his comings and goings. Her employment outside the home was also compromised. She tailored her schedule by teaching part-time so she could be free to be with Bert during his forty-eight-hour shore leaves. These types of compromises are of course similar to those in other households where one member must be away from home for long periods while engaging in paid labour such as sales, athletics, oil rig work, or the military.

In Sam and Jane's coastal fishing household, there appeared to be a single wage earner who went out to work each day and returned at night; however, Sam and Jane actually ran a family business. In this situation the domestic and public spheres overlapped. Jane had two roles: wife and business partner. She acted as the shore manager by running errands for her husband's fishing enterprise during the day and doing the business accounts on weekends, her unpaid labour making an essential contribution to the family business as well as to the household. The business enterprise functioned out of their home in other ways as well. In the winter months the basement and garage served as workshops for the production and repair of fishing gear and as storage areas. And when the fishing vessel needed to be refitted, Sam hauled it up into the backyard where he and Jane cleaned and painted it.

Coastal fishing-dependent households bear comparison with family farms, corner stores, and other family businesses. For example, tradespeople – carpenters, painters, and electricians – may operate their workshops or businesses out of their garages or basements. Women may use the family home as a daycare centre. Professionals – dentists, doctors, and lawyers – may share offices with their spouses, locate their offices in the family home, or one spouse may work as an accountant or a secretary for the other spouse. In all these cases the complementarity of home and work plays out differently in the domestic and public domains than it does in the case of wage labourers.

The Study

This study focuses on the households of coastal fishermen who lived in Lunenburg and Halifax counties, and on the households of deep-sea fishermen who sailed out of Lunenburg, the largest Nova Scotian deep-sea fishing port (see Figure 1.2). In addition to the fishing industry, the Lunenburg/Bridgewater region offered employment in tourism, light industry, and the service sector, while Halifax, the largest metropolitan centre in Atlantic Canada and less than an hour's drive from Lunenburg, offered employment opportunities in a wide range of industries. I chose this area for three reasons. First, both coastal and deep-sea fishermen and their families lived and worked in the area. Second, women could find employment outside of the fishery. Third, although the 1990s fisheries crisis undercut productivity in the groundfisheries, people could still make a living fishing. Though the combination of these characteristics existed throughout southwest and southeast Nova Scotia, they did not represent most of Atlantic Canada, where people usually depended on one employer and one market, and could no longer make a living by fishing alone.

The study has three objectives. First, by contrasting examples, the study shows, at the individual level, how men's work depended daily on their wives' labour, and how the daily existence of these women was defined, confined, and redefined by the constraints of their husbands' work. The second objective (which overlaps to some extent with the first) describes the contribution of fishermen's wives' unpaid labour, revealing the gendered nature of the apparently male-dominated fishery, and showing how production for exchange depended on this domestic labour. Third, the study enhances our understanding of how households respond to economic and ecological crises. Although this study focuses on the North Atlantic fishery, its theoretical discussions have broader implications. The research contributes understanding to three important debates: whether unpaid domestic labour plays a critical role in economic production; whether women's labour in maritime societies sustains their households' livelihoods; and whether the restructuring of these fisheries has had an impact on women and their households. It also demonstrates that the conceptualization of 'women's work' must not be limited to subsistence and wage labour. Instead, the term's definition must be expanded to include livelihood, which in turn includes social relations, the affirmation of identity, and the ownership, management, and circulation of knowledge.

The study set up an experiment of sorts. It used a comparison of two types of fisheries to show how market activities or the process of profit-making in the formal economy depended on a range of other, unpaid, typically non-market labour in households and communities. Although fishers pursued many different types of fisheries off Nova Scotia's shores, these fisheries can be divided on the basis of employment into two groups: industrial and independent. Vertically integrated companies employed 'industrialized' fishermen; 'independent' fishermen were operators/owners or crew who worked for themselves, or for 'shares.'

This distinction – industrial and independent – roughly coincided with the division between the deep-sea and the coastal fisheries, respectively.[9] In deep-sea fishing-dependent households husbands went to sea for ten to fourteen days and returned home for two to fourteen days. In these households the domestic and public domains were separate. In the coastal fishery, the degree to which household members (usually wives) contributed to the household's fishing enterprise determined the degree to which the workplace and home, and workmates and household members, coincided. In these households, the overlapping of the public and domestic domains varied from household to household, from fishing season to fishing season, from one year to the next. The degree of separation between the two domains continually changed.

Whether or not a couple could cope with the pressures associated with this way of life depended on a number of factors. These included social origins and family traditions, the stage in the life cycle of the household, locality of residence, family and other social networks, and the availability of employment opportunities and social services. The study focuses on these factors as well as on independence, the perception of independence, social attitudes, and household finances.

The characteristics of working conditions in each fishery set up inherent conflicts between spouses' adaptations. In the case of the deep-sea fishing-dependent households, the factors that facilitated women's adaptations to their husbands' absence conflicted with their adaptations to their husbands' presence. Their husbands had similar difficulties in reconciling the differences between life at sea and life onshore. Women attempted to integrate their husbands into the domestic domain and shore life; the more successful they became, the fewer conflicts arose between spouses. In the case of the coastal fishing-dependent households, conflicts arose over the privileging of the fishing enterprise's needs over all other household concerns. Husbands and wives in these households negotiated, virtually daily, the degree to which their domains of influence and control overlapped.

The project had four phases. I collected data using a range of methods, each method eliciting separate but complementary information. Phase 1 of the study consisted of a literature review and a series of in-depth interviews with selected key informants,[10] including a number of wives of deep-sea and coastal fishermen, fishermen's wives' support groups, community support workers, fish company managers, and elected officials in Lunenburg and other Lunenburg County communities. I used information from those sources to generate statistical profiles, demographic and socio-economic parameters, and to identify areas of residential concentration for the populations. Relying on the preliminary information obtained in this phase and from previous research, I developed the interview schedules and sampling plans for the subsequent phases.

Phase 2 consisted of a general survey administered to 150 deep-sea fishermen's wives from late August to November 1993, and 150 coastal fishermen's wives from late August 1993 to May 1994. I avoided interviewing in the summer months when deep-sea fishermen and their households take their vacations and the coastal fishery is at its height. Since each survey could stand on its own I analysed these data separately in turn. Our survey took snapshots, one at a time, of moments in time. If at a certain moment a household income came primarily from deep-sea fishing, I identified the household as a deep-sea fishing-dependent household; if at that moment the household income came primarily from coastal fishing, I identified the household as a coastal fishing-dependent household. Some households had moved from one fishery to the other, while others had depended their whole working lives on only one fishery.

During Phase 3, I conducted twenty-five taped, in-depth interviews with wives from each group between July and October 1994. The sample was representative regarding age, employment status, and parity (having children or not). I analysed data from each group separately. Phase 4 consisted of writing up the results from the surveys and interviews.

I handled information gathered from all sources in a manner that would guarantee the respondents' right to privacy. The interviewer explained the nature of the study in detail before beginning an interview and asked each participant to sign a consent form. At every stage of the study I ensured anonymity and privacy. Once results had been generated I sent a summary back to each of the respondents. In the text, I numbered the interviews from one to fifty, the first twenty-five from the deep-sea wives sample and the second twenty-five from the coastal wives sample. I have arbitrarily assigned names to each woman interviewed, and to her family members. That is, I called all persons in interviews 1 and 26 by names beginning with the letter *a*, all persons in interviews 2 and 27 by names beginning with the letter *b*, an so on through the alphabet, excluding the letter *x*. In a few cases, characteristics of individual households have been changed to protect the identity of the participants (see the appendix).

Set Adrift is loosely grouped into three parts. The first three chapters introduce the project and compare the basic characteristics of coastal and deep-sea fishing-dependent households. The next three chapters examine social relations, including domestic labour and the leisure activities of these households. The remaining three chapters conclude the study with a discussion concerning the financial resources available to these households, how they use these resources, and their economic and social future.

Table 1.1 presents brief demographic profiles of the two samples. In the coastal sample women's median age was thirty-eight years – 20 per cent were thirty or younger, and 17 per cent were fifty or older. Ten per cent had grade eight or less, 29 per cent had some high school, a quarter had finished high school, and 36 per cent had some post-secondary schooling. Thirty-six per cent of couples had been married for ten years or less, and a third for more than twenty years–the median length of marriage was fourteen years. Most married in their late teens and early twenties. For 10 per cent of these women this was a second marriage. On average these women had two children – 7 per cent had no children, and 12 per cent had four or more. Of those women employed outside the home (47% of the total sample) – 46 per cent had a full-time and year-round job, 69 per cent worked during the day, and 31 per cent worked split or night shifts.

In the deep-sea sample, the women's median age was thirty-nine years – 15 per cent were thirty or younger, and 12 per cent were fifty or older. Twelve per cent had grade eight or less, a third had some high school,

Table 1.1 Demographic Profiles of Samples (in percentages)

Demographic parameter	Coastal fishermen's wives	Deep-sea fishermen's wives
Age		
Median	38 years	39 years
30 years or less	20%	15%
31–49 years	63%	73%
50 years or more	17%	12%
Education		
Grade 8 or less	10%	12%
Some high school	29%	34%
Finished high school	25%	14%
Some post-secondary	36%	40%
Length of marriage		
Median	14 years*	15 years**
10 years or less	36%	30%
11– 20 years	31%	37%
More than 20 years	33%	33%
Second marriage	10%	17%
Number of children		
Mean	2*	2**
No children	7%	7%
1–3 children	81%	82%
4 or more children	12%	11%
Employed		
Proportion employed	47%	46%
Full time (35hrs plus)	56.1%	61.8%
Household's reported income		
Median	$30,000–$39,999	$60,000–$69,999
less than $20,000	13.4%	0%
$20,000–$29,999	15.6%	7.5%
$30,000–$39,999	29.6%	6.8%
$40,000–$49,999	14.1%	19.9%
$50,000–$59,999	11.9%	18.5%
$60,000–$69,999	9.6%	12.3%
$70,000–$79,999	3.7%	14.4%
$80,000 or more	2.1%	20.5%

Missing values *(4) **(2)

14 per cent had finished high school, and 40 per cent had some postsecondary school. Thirty per cent of couples had been married for ten years or less, and a third for more than twenty years – the median length of marriage was fifteen years. Most married in their late teens and early twenties. For 17 per cent of those women, this was a second marriage. On average these wives had two children – 7 per cent had no children, and 11 per cent had four or more. Of those wives employed outside the home (46% of the total sample), 46 per cent worked full time and year round, two-thirds worked during the day, and one-third worked split shifts.

In related statistical data, approximately 70 per cent of these coastal fishermen were captains who owned and operated their boats. Twenty-five per cent worked as deckhands and 5 per cent were captains on boats owned by others. They fished primarily for groundfish, lobster, and scallops. Twelve per cent of the captains had no crew, 37 per cent had one crewman, 28 per cent had crews of two, 13 per cent had crews of three, and 10 per cent had crews of four or more. Of the crew interviewed, most had worked for their current skipper for two years or less, and only 9 per cent worked for other family members. Over 50 per cent of captains reported at least one family member working for them. Of boat owners, 10 per cent had owned their boat for less than five years, while another 10 per cent had owned theirs for thirty-five years or more. The median ownership time was fifteen years. Less than one-third of these men worked year round. Depending on the fishery, they spent from one to four days at sea per voyage, alternating with overnight to four days on shore. In 1992, a quarter of these men were reported to have made less than $20,000, and 10 per cent were reported to have made more than $40,000; the median reported income fell between $20,000 and $29,999.[11]

These deep-sea fishermen were employed as deckhands (57%), officers other than captains (32%) – the latter including mates, bosons, cooks, and engineers – and captains (11%). Vertically integrated companies owned the scallop draggers, groundfish trawlers, herring seiners, clammers, and shrimpers on which these men worked, and the plants which processed their catches. Over 80 per cent of these men worked year round. Depending on the fishery, they averaged from ten to fourteen days at sea per voyage, alternating with two to four days on shore. In 1992, 10 per cent of these men were reported to have made less than $30,000, and 10 per cent more than $70,000; the median reported income fell between $50,000 and $60,000.

The demographic characteristics – women's age, education, length of marriage, and number of children – are similar for the two samples. However, deep-sea fishermen's wives were shown to be more likely to be separated or divorced and remarried than their coastal counterparts, they were also more likely to be employed full time. The median household income for coastal fishing-dependent households was half that of deep-sea households. Thus, when comparing the deep-sea and coastal households as families and as couples, the similarities are much more significant than the differences. But the differences in the samples derive directly from the variations in each household's livelihood strategies and the stresses associated with them. The next two chapters will explore these differences in greater detail.

CHAPTER TWO

Living the Dream

I made up my mind when I married him that there was two ways to go. There was to pay no attention to it and just live two separate lives as far as his work went, or to get involved and make it my life, too. And I chose to get involved and make it my life. Therefore I understand it and what's going on. (Trudy)

In the coastal fishery, the fishing enterprise was based within the household (Andersen 1979; Andersen and Wadel 1972; Faris 1966; Stiles 1971). As Trudy has indicated above, in this fishery men's and women's work spheres overlapped to a greater or lesser degree. The men ran the fishing enterprises while the women ran the households, took care of the children, and in varying degrees directly supported the household's fishing enterprise through their labour. In times of financial difficulty women also frequently engaged in wage labour to augment the household's income and support the fishing enterprise. Throughout their marriages husbands and wives continually negotiated and renegotiated the ways and the degrees to which their work spheres would overlap.

Wives' labour was crucial to the profitability of these fishing enterprises. Few coastal fishermen remained single, and those who did usually worked as helpers on other men's boats. Men who had their own vessels found it difficult to make a living if they did not have access to some form of female labour onshore. Single men and widowers relied on their mothers, sisters, and daughters to maintain their households and perform other forms of domestic labour, such as cooking meals and washing their work clothes. Many of these men also relied on female relatives for additional help – paid or unpaid labour – in the fishing enterprise itself. Without access to female labour based on kinship, these men would have had to pay others to do these jobs, thus dramatically increasing costs and decreasing profitability.

Dreaming of One's Own Boat

In the past, the coastal fishery directly recruited the sons of fishermen to replace their fathers. From an early age these boys fished with their fathers, grandfathers, brothers, or uncles, worked as helpers, and gradually took over family members' fishing boats and licences. Onshore their mothers, grandmothers, sisters, aunts, and ultimately their wives supported the fishing enterprises by performing tasks such as preparing bait, mending gear, and processing and selling the catch (Antler 1982; Faris 1966; Nadel-Klien and Davis 1988; Thompson 1985). As boats and the ever-diminishing numbers of fishing licences rose in price, and as older men stayed longer and longer in the fishery, most young men, if they wanted to enter the coastal fishery, could not expect to 'inherit' a family boat and fishing enterprise. In order to make enough money to buy a number of licences, appropriate fishing gear, and put a payment down on a boat, they had to get a high-paying job such as deep-sea fishing, work on oil rigs on land or at sea, construction work, or work in the woods. The ideal was to buy a boat fully equipped with electronics, gear, and licences, and to have a little additional money put aside before quitting your current job and entering the fishery. Many couples involved in the coastal fishery began their married lives when they were engaged in other lines of work. Making this transition was not easy. The following examples of two couples – Oriel and Obadiah, and Gilda and Graham – illustrate this point.

Obadiah and Oriel had been married eighteen years and had one child. When they were newlyweds, Obadiah worked in construction and the couple travelled together daily to and from their respective jobs. Oriel liked their time together and the type of life they had, but Obadiah wasn't happy; he had always dreamed of fishing. So he went up north to earn enough money to buy a boat, leaving Oriel behind. Although Oriel would have preferred a different type of life, she respected her husband and supported him in his desire. As Oriel said, 'I'd much rather live with somebody who's happy and just have them at certain times of my life, than to have them all the time and them be wishing to do something that they always wanted to do.'

Graham and Gilda had been married for over nineteen years and had two children. For the first thirteen years of their marriage Graham worked in the deep-sea fishery, but for the past seven years he had been gill-netting in the coastal fishery. When they first married, Graham would be gone to work in the deep-sea fishery for twelve or fourteen days at a time, leaving Gilda home on her own with full responsibility for the household. When Graham switched to the coastal fishery, he could return home most evenings and he stayed home throughout the winter. Gilda found this change 'a big, big rearrangement in life. The first year was terrible because you're used to him being gone, and then all of a sudden he's home and you're thinking, go, go somewhere, right? Because for all these years you have to break away from each other, and then he's here. You can't fish all the time. Sometimes the weather's bad [or] it's different seasons, like December's lobster season, and then you hardly do anything January, February, or March because it's too windy, too cold, and you can't really do anything, and they're there in your face, day and night (laughter).'

In these examples, wives indicated how they supported their husbands' desire to be a fisherman, compromised their own ambitions, interests and concerns to make this possible, and then had to adapt to their new circumstances. These changes included learning to cope with their husbands' earlier occupations and participation in the savings and investment strategies necessary to set up their husbands' fishing enterprises. It also meant in some cases supporting a goal that they did not believe was in their own or their families' best interest. Without their wives' financial help and other forms of support, and their wives willingness to compromise their own desires, these men could not have realized their dream.

Achieving the Dream

Look at any tourist brochure of Nova Scotia and you are likely to see a picture of a 'sleepy' fishing village situated on a harbour, with wooden wharves stretching out from the coast like fingers into the sea. Brightly painted fishing boats wait to be loaded with lobster traps and multi-coloured buoys are piled on the wharves. Unfortunately this picturesque representation of the coastal fleet belies the harsh reality.

The Nova Scotia coastal fleet consisted of vessels made of wood, fibreglass, and/or cement, all under sixty-five feet but most between thirty-five and forty-five feet in length. They plied the coastal waters usually no more than fifteen nautical miles offshore. Individual households owned and operated these boats. In our sample, 70 per cent of coastal households owned their fishing business. One man operated his own boat during lobster season but worked the rest of the year as a captain for a deep-sea fishing company. The remaining 30 per cent of the men had jobs as 'helpers' or crew.

These enterprises, based in individual households, had three major business components: the harvesting of marine resources at sea; support services on shore; and financial management, including the sale of the catch. These businesses ran on a combination of household and non-household labour, and on paid and unpaid labour. Historically, coastal fishing-dependent households engaged in a combination of commodity production, wage labour, and unpaid subsistence work – segregated strictly along gender lines (Antler 1977, 1982; Davis 1983a,b; Porter 1983, 1985a,b; Thompson 1985). Men fished, built and maintained their boats, made and repaired their gear, hunted, engaged in wage labour during the 'off-season,' and maintained the physical structure of their homes and outbuildings. Women kept house, raised the children, baited trawl, helped with repairs, processed the fish, maintained the accounts, and tended gardens and domestic animals. Few women fished (Antler 1982; Porter 1983, 1985b).

Since the 1960s this system had broken down, although this gendered division of labour continued to some extent into the 1990s (Davis 1983b; Porter 1983; Sinclair and Felt 1992). Women no longer processed the fish within the household. Some households sold fish directly to brokers for the fresh fish market in Canada and the United States. Others sold their fish to industrialized fish plants where women, and to a lesser extent men, processed the fish on assembly lines (Ilcan 1986). But women continued to work as shore crew and maintained the financial accounts of household enterprises. A few women went to sea as helpers (Sinclair and Felt 1992; Thiessen, Davis, and Jentoft 1992).

In the 1990s coastal fishing enterprises attempted to spread the economic risk over a number of fisheries. They harvested valuable species such as tuna, crab, lobster, and scallops, as well as underutilized species such as shark, dogfish, mackerel, surf clams, and some groundfish, but at reduced levels. These marine resources were harvested throughout an annual cycle according to their seasonal availability. Thus the successful coastal fisherman held a variety of licences and harvested these seasonal resources when he found it most advantageous to do so. But each fishery had a boom and bust cycle. In the 1990s the most lucrative marine resources – lobster, crab and scallops – took on increased importance with the decline in the groundfish stocks. Fishermen tended to exploit the most profitable fishery and then move on to another species when the first fishery had been closed for the season, or if the resource went into decline (Apostle et al. 1998).

Work patterns and the organizations of labour both at sea and onshore varied depending on the fisheries being exploited. Each fishery had different fishing seasons requiring different types of gear, work patterns, fishing schedules and crew sizes, as well as different forms of support from the fishing enterprises' households. Crew sizes ranged from one (e.g., lobstering) to four (e.g., some groundfishing and scalloping). The length of the voyage ranged from day sails (e.g., lobstering) to a week (e.g., swordfishing). Some types of fishing required more household labour (e.g., salting cod) and others less (e.g., acting as shore crew for swordfishers). All coastal fishermen in our sample took some time off during the year, usually for boat and gear maintenance, family vacations, hunting, illness or accident, and Christmas. Nearly 90 per cent of these men collected unemployment insurance[1] sometime over the year prior to the interviews.

Each coastal fishing-dependent household in this study had developed a personal strategy for exploiting the marine resources. By looking at two specific households – Brenda and Bruce, and Gilda and Graham – we can see how the activities of both the household and the fishing enterprise changed throughout the year; how these strategies needed to take into account the demands of women's work inside the home, the family's stage within the life course, and employment outside the home; how women modified their work to accommodate these demands; and how each household's fishing enterprise depended on women's labour.

Bruce and Brenda had been married for eighteen years and had two young children. Throughout their marriage they developed a division of labour that met both the demands of their fishery enterprise and those of their domestic household. Bruce, primarily a lobster fisherman, supplemented his income by swordfishing. In the study area the lobster season ran from the first Monday in November to the end of May. During the off season – June through November – Bruce maintained his boat and home, and worked in other fisheries.

Lobstering is labour-intensive. When lobstering, fishermen normally work in pairs using lobster pots or traps that need to be hauled and baited daily, and they usually return home every night. The catch must be kept alive, and should be sold as soon as it comes ashore. Bruce paid Brenda a salary for six months' work, two months of which involved mostly maintenance jobs including repairing and painting the boat and the 250 lobster traps and buoys. During the lobster season, Bruce and his helper worked on the boat and Brenda acted as the shore manager. Brenda's job then consisted of 'going for bait and bringing it home [and] salting it. When he fishes I have to be available because he can be out there and say, "Listen, I need a part for my boat the next day." Well, I have to jump in the car and get the part because he'll get in at dark and go the next morning when it's dark. I have to do all the running around when he starts lobster fishing because that's where he makes his big money. So if he needs anything for the maintenance of the boat, I have to do it. Lots of times I will bail [put rubber bands on the claws] the lobsters. Banking, everything. I have to do all the finances.' During the 'off season' Bruce swordfished with one or two other men, worked on his boat, and collected unemployment insurance when not fishing. When swordfishing, Bruce remained away from home for two or three weeks and Brenda did not work as shore manager. Upon his return, her contribution then consisted of domestic service such as washing his work clothes, preparing his meals, and occasionally preparing meals for the crew.

Gilda and Graham, introduced earlier, had been married for almost twenty years and had two children in their late teens. They employed a different strategy from that of Brenda and Bruce. Primarily a gill-netter/longline fisherman, Graham supplemented his income by lobster fishing only during the most lucrative month, December. Each year he hoped that lobstering would be good enough to clear up any outstanding bills he had on his boat, because he knew his groundfishing would not. After lobster season, from January through March/April, he usually stayed home making nets, repairing gear, and fixing up the boat; then he fished for groundfish from April to October, depending on the weather. In November he readied his gear for the lobster season.

Graham usually remained at sea for one or two days when gill-netting or longlining – two fishing methods used to harvest groundfish. Gill-netting consists of suspending a net in the water, which catches fish by their gills when they attempt to swim through it. Longlining refers to fishing with baited hooks placed at intervals along lines hundreds of feet long. After each hook is baited, the lines are stored in large tubs on shore until loaded onto the fishing vessel. When the fishermen reach the fishing grounds, they play the baited lines out into the ocean where they either float, or sink and become anchored at sea bottom. Both types of fishing require two- or three-person crews, including a captain, to set, check, and haul the nets or lines. Longlining usually results in a better quality of fish, but the process requires more labour and costs more than gill-netting because each hook must be baited and unhooked by hand.

Gilda worked as shore manager during the groundfish season, but without formal remuneration. Describing her contributions to the family's fishery enterprise, she said, 'When it's gill-netting my duties are: buy the boat groceries, order the fuel, pay bills to any suppliers for salt, repairs, maintenance, or whatever, organize him because he's very unorganized (laughter), keep a record of his books for income tax, for GST. [At the] end of the year you have to put everything in piles and separate it from pay stubs to this, this, this, and this, to take to an accountant to do your income tax. I used to bait trawl before I went to work for another employer to have a paycheque coming in. [I] clean the building, clean the basement. That's about it, I guess – generally everything.' For the year prior to the study, Gilda had also been employed full time doing shift work at the local factory.

In both of these examples, the livelihood strategy used by each household incorporated women's labour – both paid and unpaid – while accommodating the specific circumstances of the fishing enterprise, the family's stage in the life course, and labour force participation. Unlike Gilda, Brenda's involvement in the lobster fishery coincided with her 'free time.' Brenda worked for pay as shore manager for the family business during the lobster season, when her children were in school and her family responsibilities were less demanding. With two young children, Brenda's busiest time occurred in summer when Bruce groundfished and did not need her labour in the fishing enterprise. This arrangement allowed Brenda to concentrate her attention on their children and her domestic chores during the summer without jeopardizing the profitability of the fishing enterprise.

Gilda, on the other hand, had a forty- to forty-five hour per week job at a factory, worked for her husband's fishing enterprise, without pay, and maintained her household responsibilities. With two older children – a teenaged daughter and a son at university – Gilda did not feel compelled to stay home to look after their children during the summer months. This 'freedom' allowed her to take on paid employment while maintaining her unpaid involvement in the fishing enterprise.

Helping Out

This flexibility and accommodation of the women to the demands of both the fishing enterprise and the needs of the household characterized all coastal fishing-dependent households (Davis and Nadel-Klien 1992). As we have seen, women in these types of households not only contributed in a myriad of ways to their household enterprises but also sustained the fishery itself during times of crisis (Davis and Gerrard 2000). I asked coastal fishermen's wives about their involvement in their households' fishing enterprises.[2] The majority of replies indicated that women saw the fishing enterprise as primarily their husband's domain, one in which they 'helped out.' For example, Unice, who had been married to a coastal fisherman for thirty-six years, still saw herself as helping out: 'They would make all the decisions ... when they would go out and when they would come in. That was all their business. We just were down there to help. We didn't make decisions. If the fish was to be sold, I done a lot of calling for the prices and stuff like that. But he made the decision where it was to be sold and where our fish was going, and what [was] the best opportunity for us to get the most out of our fish and what buyer was the best one to deal with. That was all his business.'

Like Unice, many fishermen's wives felt they were only 'down there to help.' But how much labour were wives willing to give to the enterprise? At the beginning of this chapter Trudy stated that upon her marriage seven years before she had had to choose between living a life separate from or integrated with her husband, Tim. Every coastal fisherman's wife had to make this choice, and consequently women's involvement varied dramatically.

All fishermen's wives supplied their husbands with domestic services such as washing clothes and cooking meals. But as we have seen in the cases of Brenda and Gilda, wives' involvement in the fishing enterprise might go beyond domestic service to their husbands. Women's involvement might have included contributing to the business side of the enterprise: keeping the books, buying supplies, and selling the fish; working as a shore manager; and being active partners through direct participation in the enterprise, including the catching of fish. All of these tasks might have been paid or unpaid labour. Tables 2.1, 2.2, and 2.3 summarize this wide range of activities and report the involvement of coastal fishermen's wives in their husbands' fishing enterprises in the areas of direct participation, business and financial management, and support and domestic services. It also compares these results with an earlier study by Thiessen, Davis and Jentoft (1992).

In Table 2.1, the values reported for wives of coastal fishermen (column 1) indicate moderate levels of both direct and indirect participation in their households' fishing enterprises. In all except a few areas,

Table 2.1 Reported Involvement of Wives in Their Husbands' Fishing Enterprises: Comparison of Coastal Sample and Findings of Thiessen, Davis, and Jentoft (1992) (in percentages)

Tasks	Wives study coastal – all wives	Nova Scotia – captains' wives*
Direct participation		
Bait trawl/line	11	10
Repair nets or traps	12	15
Knit lobster heads	3	18
Clean and/or salt fish	17	22
Fish with husband	25	28
Business/financial management		
Arrange sales of catches	19	11
Prepare income tax	13	21
Arrange for credit	13	30
Keep the books	52	54
Keep record of catches	36	59
Pay fishing bills	53	61
Track fish prices	26	N/A
Support services		
Go for parts for the boat, etc.	47	51
Listen to CB or marine radio	48	59
Clean the boat	27	41
Domestic services		
Make meals for the crew	38	43
Clean fishing clothes	95	98
Number of cases	150**	157

Missing values ** (1)
N/A = Not asked
*Thiessen, Davis, and Jentoft (1992, 345)

these women reported lower participation rates than those found in the Thiessen, Davis, and Jentoft (1992) study (column 2) of Nova Scotian, coastal fishing households. Thiessen, Davis, and Jentoft (1992) interviewed only households where the husband owned and operated his boat, so their sample consisted entirely of 'captains' wives.' My sample includes 106 captains' wives and 44 crewmen's wives.[3] Table 2.2 compares the coastal sample broken down by job status – captain versus crew – with that of captains from the Thiessen, Davis, and Jentoft (1992)

Table 2.2 Reported Involvement of Wives in Their Husbands' Fishing Enterprises: Comparison among Coastal Wives by Husband's Job Status (in percentages)

Tasks	Wives study coastal – crews' wives	Wives study coastal – captains' wives	Nova Scotia – captains' wives*
Direct participation			
Bait trawl/line	9	12	10
Repair nets or traps	2	16	15
Knit lobster heads	0	5	18
Clean and/or salt fish	9	20	22
Fish with husband	11	31	28
Business/financial management			
Arrange sales of catches	7	24	11
Prepare income tax	14	12	21
Arrange for credit	9	15	30
Keep the books	14	68	54
Keep record of catches	18	44	59
Pay fishing bills	11	70	61
Track fish prices	7	34	N/A
Support services			
Go for parts for the boat, etc.	16	59	51
Listen to CB or marine radio	29	55	59
Clean the boat	16	33	41
Domestic services			
Make meals for the crew	18	47	43
Clean fishing clothes	95	95	98
Number of cases	44	106	157

Missing values (1)
N/A = Not asked
*Thiessen, Davis, and Jentoft (1992, 345)

study. When we only look at the values for captain's wives, these results more closely resembled those from the earlier study. As indicated in Table 2.2, column 2, the involvement of crewmen's wives was substantially less than that of captains' wives for all activities except washing fishing clothes.

A small number of women, mostly crewmen's wives, limited their involvement to domestic services. Usually these women got up at three-thirty or four in the morning, prepared their husbands' breakfast, made their lunches, sent them off to work in clean clothes, and prepared their dinners for their return. A few women, such as Anne, who had been married eleven years to Abraham, recognized the importance of their domestic labour – that it made their husbands' work possible. As Anne said, 'I basically get him ready to go again, feeding him, keeping him clean, and picking up things he needs.' These contributions to a husband's well-being resemble those of many women in blue-collar households.

On the other hand, helpers' wives could maintain some distance from their husbands' work. Claire, married about two years, was in fact disgusted with the fishery: 'I don't want nothing to do with fishing, nothing. It's gross.' If they preferred it, most helpers' wives could avoid involvement beyond 'domestic service,' but these services were still key to maintaining the fishery. The few crewmen's wives who participated directly in the fishery to supplement their household income, usually worked for their husband's boss. For example, Claire baited trawl for her husband's employer even though she found the fishery so unappealing. As she explained, 'However many times we [she and Colin] baited, it came off of Colin's expenses. So it saved us money all around. He got that much more of a pay because he didn't have to spend out so much of it on expenses.' Similarly, other crewmen's wives gutted, cleaned, and salted fish, helped with repairs, and did routine shore jobs, thereby decreasing their husbands' expenses.

As Table 2.2 shows, almost all captains' wives participated, at least indirectly, in their husbands' fishing enterprise. Sixty-eight per cent kept the books, 70 per cent paid the fishing bills, 15 per cent arranged credit, 45 per cent kept records of the catches, a third tracked fish prices, and 24 per cent arranged for the sale of the catches. Many (59%) went for parts for the boat and cooked meals for their husbands' crews. About a third cleaned the boat, and 95 per cent washed their husband's fishing clothes. Direct participation in the fishing enterprise was lower. With reference only to captains' wives, 31 per cent fished with their husbands. A few of these wives cleaned and salted fish (20%), repaired nets or traps (15%), baited trawl (16%), and knitted heads for the lobster traps (5%). However, this 'laundry list' approach obscures the interdependency and interaction between husbands and wives' work spheres, as the earlier discussion of the households of Brenda and Bruce and Gilda and Graham illustrated. This approach also does not indicate how women's activities in the fishing enterprise reorganized or restricted their other daily activities.

In these earlier examples – Brenda and Bruce, and Gilda and Graham – each household had developed a livelihood strategy that incorporated women's paid and unpaid labour while accommodating their specific personal circumstances. Thiessen, Davis, and Jentoft (1992) reported that Nova Scotian coastal fishermen wanted their wives to be more involved in their fishing enterprises than they then were. Thus every couple needed to decide how involved the wife would be, and to identify the tasks she would take on. If she did not have the skills to perform certain tasks she would have to learn them, then integrate them into her other commitments – household and family responsibilities, and possibly employment – while attempting to maintain her own autonomy. Invariably, conflicts between spouses were common. Both Thiessen, Davis, and Jentoft (1992) and Munk-Madsen (2000) reported similar conflict in Norwegian fishing-dependent households.

When women entered the workforce their involvement in the household fishing enterprises changed. For example, when I asked Gilda how her contribution changed when she started paid work, she replied, 'Ordinarily, if I wasn't working, I'd be down there baiting trawl. As far as the rest of the stuff, I still do all of that but it's very rushed.' As part of their household strategy to maintain their fishing business, Gilda had reduced one form of work – unpaid labour within the fishery – for paid employment outside the fishery. Although Graham now had to pay someone to bait trawl for him, the household's net income had increased substantially.

Table 2.3 compares the involvement of the forty-seven captains' wives who held jobs outside the home with the fifty-nine captain's wives who did not. Both groups reported virtually the same level of involvement in washing their husband's fishing clothes, listening to CB and marine radio, arranging sales of catches, keeping records of catches, tracking fish prices, and fishing with their husbands. Women who had jobs reported higher involvement in five activities: baiting trawl or lines, cleaning and/or salting fish, preparing income tax, arranging credit, and cleaning the boat. Women who were not employed reported higher involvement in the six remaining activities: repairing nets or traps, knitting lobster heads, keeping the books, paying fishing bills, going for parts for the boat, and making meals for the crew.

Two areas of increased involvement by employed wives – baiting trawl and salting and drying fish – seemed counter-intuitive, since both employment and these tasks put significant constraints on women's time. However, these labour-intensive tasks were associated with the

Table 2.3 Reported Involvement of Wives in Their Husbands' Fishing Enterprises: Comparison among Coastal Captains' Wives by Wife's Employment Status (in percentages).

Tasks	Wives study coastal – captains' wives	Wives study coastal – captains' employed wives	Wives study coastal – captains' homemakers
Direct participation			
Bait trawl/line	12	15	10
Repair nets or traps	16	13	19
Knit lobster heads	5	2	7
Clean and/or salt fish	20	25	15
Fish with husband	31	32	30
Business/financial management			
Arrange sales of catches	24	25	22
Prepare income tax	12	17	8
Arrange for credit	15	19	12
Keep the books	68	62	73
Keep record of catches	44	43	46
Pay fishing bills	70	62	76
Track fish prices	34	32	36
Support services			
Go for parts for the boat, etc.	59	55	62
Listen to CB or marine radio	55	55	55
Clean the boat	33	38	27
Domestic services			
Make meals for the crew	47	42	51
Clean fishing clothes	95	96	95
Number of cases	106	47	59

Missing values (0)

catching and processing of groundfish, the hardest-hit fishery and thus the hardest-hit households. Women like Gilda, whose households depended on the groundfish fishery for survival, engaged in paid employment to offset the difficult economic circumstances of their households. As Gilda said she took a job because 'Graham's been inshore fishing now, I think, six or seven years, and every year his income has decreased, like a lot. Without me working for the last year it would've been bad, right? Because he would've never made enough for us to survive on. So I said last summer, when there was hardly any fish, "I'm going for a job in September." "Don't be so foolish," he said, "don't worry. Don't worry about it." But I knew I had to go do something.'

In order to maximize the economic advantage for their households, women needed to continue to perform labour-intensive tasks as well as working in the paid labour force. Without this 'triple day' the household's fishing enterprise would have had to hire someone else to bait the trawl or to process the fish, resulting in a limited or negative gain in household income. This meant that women, like Gilda, whose husbands depended on groundfish and who were employed, needed to maintain their substantial additional workloads in order for the fishing enterprise to be profitable.

In the Thiessen, Davis, and Jentoft (1992) study of captains' wives, 86 per cent were employed, but in my study only 44 per cent worked in the paid labour force. The difference in these employment rates may be related to the higher proportion of groundfish fishermen in the Thiessen, Davis, and Jentoft sample. Other factors, including different fishing seasons, degree of reliance on lobster as the main fishery, degree of profitability of the fishing enterprise, and different stages in the family life course, may also have contributed to the higher involvement levels reported in the Thiessen, Jentoft, and Davis study (see Table 2.2, column 3).

Some women who participated heavily in their husbands' fishing enterprises argued that this involvement precluded them from paid employment. For example, Brenda stated that her work as shore manager made so many demands on her time that 'I can't really hold another job. He employs me, and it's best he does employ me because I handle all the finances and what not.' But for Brenda working as shore manager meant paid employment: 'He employs me.' Unlike Unice, Brenda did not 'help out' her husband, she 'worked for him.'

Thus, the family life course put various and ever-changing stresses on women's work. Coastal fishermen's wives experienced a range of pressures: newly married, coping with small children, managing school-age children, dealing with teenagers and ailing parents and grandparents. These pressures then competed with those associated with the needs of the fishing enterprise. Compromises were made in every household. For example, Brenda did not participate in the fishery when her children were home from school, but did during the fall and winter months. Gilda refrained from going to work until her youngest was in high school. Women had to be flexible to satisfy the changing and competing demands of the household and the fishing enterprise, even more so with the fisheries in crisis.

The Dream in Peril

The fisheries crisis led to a reorganization and restructuring of individual fishing enterprises, fishing–dependent households and the fishing industry itself (Apostle et al. 1998; Binkley 1996, 1995a; Neis and Williams 1997; Newell and Ommer 1999; Sinclair and Felt 1992). In all cases, incomes of coastal fishermen in this study dropped dramatically, but coastal fishing-dependent households adopted several strategies to offset growing costs. In response to the restrictions on catches and the shortage of fish, coastal fishermen developed four main strategies, which could be used individually or in tandem, to meet the crisis: (1) remain at sea longer and exploit resources farther offshore or in richer fishing grounds; (2) exploit as many species as possible; (3) employ as much household or family labour as possible; and (4) use income from household members' wage labour to subsidize the fishing enterprise. Each strategy exerted different forms of tension on current household adaptations to the coastal fishery. A few fishermen rejected the trade-offs involved with each of these strategies and opted instead to continue to exploit their traditional fishing grounds while increasing their fishing efforts.

There had always been a tradition within the coastal fishery of household members' seeking employment outside the fishery when things got financially tough. As discussed earlier, in times of financial crisis fishing-dependent households relied on income from women's paid employment to carry them through. When this stratagem proved inadequate – that is, when wages from wives' part-time or full-time jobs could not meet the needs of the household – men then considered seeking employment outside the fishery. For example, they sought employment during the 'off season' and fished only the most lucrative fisheries, such as lobster, or mothballed their vessels and worked at other jobs year-round until the boat and other expenses were paid off. In the best of times, realizing the dream of owning one's own boat, purchasing a home, and raising a family put a huge financial strain on a couple. Most fishing-dependent couples set priorities and decided how much they would sacrifice to keep the boat running. In many cases the arguments concerning the economic strain that boat expenses put upon the household budget became the major cause of friction between husband and wife. In economic hard times the level of tension in many households increased. The two basic questions, then, for these men and their households were: Are the monetary benefits sufficient? and Is the erosion of the non-monetary benefits (e.g., the long time away from home or less safe working conditions) worth it? Four couples – Brenda and Bruce, Gilda and Graham, Stewart and Susan, and Laurie and Len – provide examples of using combinations of these strategies to cope with the crisis in the coastal fishery. (Earlier in the chapter we compared the fishing enterprises of Gilda and Graham, who primarily harvested groundfish but maintained a lobster licence, and Brenda and Bruce, who participated primarily in the lobster fishery but also swordfished.)

As mentioned earlier, Bruce and Brenda held a number of fishing licences. With the groundfish stocks severely reduced, many coastal fishermen retained fishing licences for future exploitation, but came to rely on other fisheries – primarily lobsters – for their income. But lobster stocks, like other marine resources, go through cycles, and there might be nothing to fall back on in the future. In order to be flexible, fishing enterprises must maintain other fishing licences. As Brenda explained: 'Bruce just mainly depends on lobstering, but he has other fishing licences. He handlines, swordfishes, and things like that. If anything goes wrong with the lobster season, with the catch of the lobsters, sometimes that runs in cycles, we've got nothing else to fall back on. There isn't any other fish out there to catch. And he's got a $75,000 boat sitting down there, and a $20,000 licence, piece of paper, or whatever. And if the lobsters go, if he wanted to get out of the fisheries, the way the fishing is we could never get the money that he put into it.'

The real economic problem lay in the reliance on one species as the mainstay of a household economy. The fisherman who could afford to hold a number of licences could spread the risk: if one fishery declined then another could be exploited. All fisheries followed this cyclical pattern, and few fisheries in the North Atlantic rebounded as quickly to baseline stock levels as they did in the past. Fishermen recognized this cyclical nature and the need of fish stocks to rebound in the future. Brenda continued: 'Our income – we depend and live off just lobstering. So that's been okay. If you come [to] interview [me] maybe in another three or four years, I might be pretty sad looking. I'll have a nine-to-five job somewhere, and I'll be the working spouse (laughter). Lobstering took a cycle one time before; Bruce just got into it when the boom came back up. But I think we can look for a cycle, another few years that we won't be able to live on lobstering. I'm thinking about food and bills, not just all the frills, right? But I don't know, really, like the other fish aren't going to come back.'

Another concern was the market. Many fishermen feared that the price of lobsters would fall. Since they were putting more effort into catching each lobster, they wanted to be rewarded for this effort with at least the same profit level as before. Thus another cycle developed that revolved around rising prices, consumer demand, increased fishing effort, and declining profit levels, which placed additional strains on the resource.

Since lobster and groundfish harvesting were labour-intensive, another way to cut costs was to have not only wives, but also sons or other family members participate directly in the fishery, either as helpers on board or as shore crew baiting lines, repairing nets, gear, and traps, or cleaning and salting fish. Consolidation of family labour in order to maintain profits within the fishing-dependent household was also a common strategy used by Norwegian households during their fisheries crisis (Munk-Madsen 2000). This strategy had frequently been used in the past in Nova Scotian fisheries where fishermen returned home most evenings. But even in these situations some fishermen still could not make a living. Many of the women spoke about how the amount of work they did in the fishery had dramatically increased, particularly those chores associated with the lobster fishery or the salt-fish trade.

The viability of Gilda and Graham's fishing enterprise appeared to be less secure than that of Brenda and Bruce. Though part of their livelihood strategy involved Gilda's working outside the fishery for wages to augment Graham's income from fishing, Graham fished further and further away from home, and stayed out longer and longer at sea – occasionally for weeks at a time. These long absences away from home not only put emotional strain on household members, they eroded wives' involvement in the household fishing enterprise while increasing wives' household responsibilities.

Longer fishing trips associated with the groundfishery were particularly non-conducive to wives' working as helpers on board fishing boats because the work was incompatible with many family and household commitments. Even women whose work was restricted to shore-based activities (e.g., baiting trawl, selling catches) had to cease their contributions because of the distances involved. In order to sustain economies, husbands themselves took on most of the tasks previously done by their wives, while the wives felt under pressure to substitute their former unpaid labour on their husbands' boat with paid employment.

For a number of coastal fishermen's wives, the intensification of fishing meant that for the first time they had to cope with running their households completely on their own, which required them to develop the independence needed for the task. For some, like Gilda, who had a full-time job and two older children, this was an opportunity to be seized. It was also a return to an earlier stage of her married life when Graham worked in the deep-sea fishery: 'See I'm independent,' she said. 'Some people, oh they don't want him to go or what not. I'm not that personality. I'm not scared to stay alone. It doesn't bother me.' For other women, especially those with small children, this change created increased responsibility, additional work inside the home, and added stress. Many of the comments made by coastal fishermen's wives resembled those made by deep-sea fishermen's wives concerning the challenges associated with their husbands' extended time at sea.

Stewart and Susan had been married for seven years and had four young children. They had maintained their old fishing practices, and had supplemented their household income directly through Susan's wage labour outside the fishery and indirectly through Stewart's increased participation in household tasks, including childcare. Stewart, a gill-netter, continued to fish out of a port close to home rather than go further away to Georges Bank, a richer fishing area. Susan worked full time, but only seasonally, in retail sales, but her income made it possible for Stewart to fish near their home port. Susan saw their situation as a series of trade-offs. Stewart's boat had modest but consistent catches. His expenses remained low, partly because Susan could 'help out' working as shore manager when necessary. Much of Susan's modest income went directly to pay for the household's expenses; Stewart looked after the children when Susan went to work, thus offsetting the cost of a babysitter. This arrangement also allowed them a home life, which both of them craved. But Susan's seasonal job began each May and ended each September, so for the household to remain financially viable they needed to cut expenses or increase revenues during the rest of the year.

Susan and Stewart negotiated an increase in her involvement in the fishing enterprise. As Susan explained:

> I told him this fall when school starts and I'm finished this job that I would be willing to help him more if that meant cutting the bait up for them to use when they're baiting. So I'd like to learn because the gentlemen that's

with him now can only do two tubs in a day. Well Stewart can do twelve. So I said, 'That's not fair, and the guy's getting the same pay.' Then he hired people to come and bait trawl, and that's fifteen dollars a tub, and so far in four weeks we've lost, well Stewart and his brother lost, four thousand dollars baiting money out for baiters. And I said, 'That's ridiculous – like two thousand dollars of that could come on our account. So I said that when I'm finished this job that I'm going to do my best to help him.

By taking on this additional work, Susan would not only decrease her husband's expenses and compensate for her lost wages, but she would also learn a valuable skill that she could use if things got even tougher. For Stewart and Susan this was the right balance, but few households were willing to make the economic sacrifices and take the risk of smaller and less lucrative catches that these strategies entail.

The last couple, Laurie and Len, who had been married for eight years and were expecting their first child, had virtually opted to drop out of the coastal fishery except for the short lobster season each December. For the first five years of their marriage Len worked on his own coastal fishing boat, but as Laurie explained, they encountered some problems.

> Everything we had went into it [their boat]. The idea of owning your own boat is a real dream, but it's a lot of responsibility. Every headache is yours. Where with a company, if something doesn't go your way you come in and you hand in a list saying what has to be done, and you go home, or you should be able to go home. You can't do that when it's your own. We were together eight years, so for the first five years, off and on, he fished his own boat and we fought non-stop. We really had a hard time, in my opinion, and I think he would agree. I was working full time, and I'd say probably 75 per cent of the money I was clearing was going into his boat. Finally ... we had to put a hold on that.

So in order to keep their boat and their marriage intact, Len began working year-round as captain on a deep-sea vessel for a company based in Cape Breton. (Len got the job because prior to buying his own boat the company had employed him, and he held a Master Mariner I fishing licence.) Len spent ten days at sea and had a turnaround time for the boat of eighteen to twenty-four hours. During the lobster season, Len took time off from the deep-sea fishery and went lobstering on his own boat.

With this schedule Laurie and Len found it difficult to make time for each other. At the time of the interview, Laurie had quit her job in order to be available when Len came ashore, especially in anticipation of the birth of their first child. Moreover, the deep-sea fishery was more lucrative. They no longer relied on Laurie's wages to pay for the boat and support the household fishing enterprise. Still, Laurie found it difficult to have quality time with Len. As she remarked:

> Well, there's time that we have to make exceptions. We have to sit down and say now whoa, where do I belong here, where do I fit in, because I find Len gets tunnel vision. He has a one-track mind: boat, boat, boat. So whoa, where do I fit in? That's when it's usually just a day put aside that's no boat, just me, just us. Not very often, maybe three times a year. See, where I'm not working I can go where the boat is. Now we're going to be having a family in the spring, so what we're going to do then we have no idea – if we're going to just strap her to our backs and take her with us (laughter). We have no idea how we're going to handle it, but we'll go with the flow.

Laurie and Len traded off time with each other for more financial security. Much of their life resembled that of a deep-sea fishing household, though Len came home every night during lobster season. Although financially tight since Laurie left her job, they made the payments on their boat, car, and house. Laurie did not know where their future lay, but she wondered if Len would ever leave the deep-sea fishery for good and return to work on his own boat full time.

As the example concerning Len suggests, coastal fishermen often saw their self-esteem and self-worth as defined by their ability to fish; thus, even the thought of giving up fishing was difficult. Unlike the deep-sea fisherman, whose employment could be terminated or reduced at any time by the company, the coastal fisherman had to make the decision to quit himself or be closed down by the government through fishery regulations. Some women said their husbands would fish no matter what, or until the government closed down all of the fisheries; others said their husbands would wait patiently until the fishery re-opened to start again. Still others reported that their husbands spoke about possibly leaving the fishery because of the mental and emotional stress associated with economic hardship. However, none of the woman I interviewed said that their coastal fishing husbands would leave the fishery voluntarily.

For young couples just getting into the fishery, the decision to stay or to go involved a number of factors. For those with little invested, both physically and emotionally, taking the TAGS package posed a serious possibility. But for others, whose dreams of the future involved having their own boat and taking over the family business, especially where their extended family had too much invested in the enterprise, this possibility was no more than hypothetical. In some cases, the fisheries crisis hastened the departure of older family members from the fishery, which enabled sons and other younger male relatives to take over the family businesses earlier than they might have expected. But there was still no guarantee for the future.

Still, for many fishermen, the economic considerations remained secondary to their desire to fish and to be at sea. It seemed, as well, that the longer these men fished the more difficult it was for them to give it up. The key for remaining in the fishery appeared to be alternative sources of household income that allowed the household fishing enterprise to limp along until the stocks rebounded. The assumption that wives would take on or maintain a job in order to help support the fishing enterprise underlay most households strategies. Women's labour was seen as a household resource that could be used either by the household or in the workforce to offset economic difficulties caused by the fisheries crisis, thus subsidizing the fishing industry. In the best of economic times, female labour – both paid and unpaid – supported the fishery, but in times of financial distress it sustained the fishery.

Throughout this chapter, we have seen how women's labour was crucial to the day-to-day operation of the household-based fishing enterprise. Restructuring of the fishery made women's labour, both paid and unpaid, within and without the enterprise, even more important to the sustainability of the fishing industry. Daily, within these households, men and women negotiated the gendered division of labour. Women willingly took on these additional burdens because they recognized the needs of other household members, acknowledged them as legitimate needs, and took actions to fulfil them. These actions reflected a 'traditional' female response that was essentially caregiving: by taking on these additional burdens, women reasserted their femininity (Benjamin 1988). Thus women's social and emotional support, as well as their labour, sustained the coastal fishery. In the next chapter, we will explore how women's labour, and their social and emotional support, sustained the deep-sea fishery as well, and how the fisheries crisis modified the division of labour in those fishing-dependent households.

CHAPTER THREE

Two Separate Worlds

You have to realize what you're getting yourself into. You're marrying a fisherman. You have to know what that means. You have to know that he's not going to be home all the time. There are going to be some really hard times in your life when he's just not going to be there. (Paula's mother)

In the deep-sea fishery, fishing enterprises were dominated by large international vertically integrated companies which resembled other industrial settings where men's and women's work spheres remained virtually separate (Binkley 1995a,b). However, the nature of deep-sea fishermen's work, directly and indirectly, compromised their wives' domestic labour and paid employment (Binkley and Thiessen 1988; Danowski 1980; Thompson 1985). The organization and type of work done by deep-sea fishermen required that these men be at sea for prolonged periods with only a few days ashore before returning to the sea. This defining feature of deep-sea fishermen's employment meant that women had to take on their husbands' household responsibilities and obligations while the men fished, so that they could devote their full attention and energy to their work. Wives then modified their domestic labour and paid employment routines to accommodate their husbands when they returned to shore.

In these fishing-dependent households the constraints of men's work defined, confined, and redefined the nature of women's work, both paid and unpaid. If women had not been willing to make these compromises or to take on additional responsibilities and obligations, their husbands would not be employed in the deep-sea fishery. Moreover, these women, through their roles as caregivers, mediated the effects upon

44 Set Adrift

their husbands of the harsh working conditions which characterized the industry, allowing these men to continue their work. The labour of these women was just as crucial to the deep-sea fishing industry as the labour of the wives of coastal fishermen was to their industry. Throughout this chapter we will see how the interdependency between the work domains of deep-sea fishermen and their wives played out in everyday life, and how this interdependency supported the fishing industry.

His World

In Nova Scotia's deep-sea fishery, men worked for one of eleven vertically integrated companies based in the Atlantic region, harvesting scallops, groundfish, shrimp, surf clams, and other commercial species (Binkley 1995b). In steel-hulled vessels, 65 to 150 feet in length, these men sailed from their home ports in Nova Scotia, south to the Georges Bank, north to the Davis Strait, west into the Gulf of St Lawrence, and east to the far reaches of the Grand Banks. The companies recruited young men, generally with little experience outside this fishery, whom they hoped to retain within the industry for their entire productive working lives. These companies, supported by the federal government through the Department of Fisheries and Oceans (DFO) and by the provincial government through the Department of Fisheries, promoted training programs and the professionalization of their workers in order to maintain a well-trained and well-paid workforce with little employment turnover.[1] Once a fisherman joined the deep-sea fleet his work schedule and the concomitant demands on his household remained relatively constant throughout his career. Specific working conditions and schedules varied with the type of marine resources harvested and the fisherman's career cycle, but work in all sectors of the fishery involved long absences from home.

Companies organized fishermen employed on these vessels into a managerial pyramid reflecting an industrial model. The master or captain oversaw the running of the vessel with the aid of the other officers – first mates, engineers, bosuns, and cooks. The crew, which included trawlermen, deckhands, and learners, formed the base of the pyramid. (The basic social division between captains and officers, and the crew, continued ashore.) Each job on board had a classification according to defined activities and a graded pay distribution. Captains and first mates, holding supervisory positions, managed the harvesting of marine resources at sea. They also exercised their traditional roles as persons responsible for finding the fish and navigating the vessels. In this study 77 per cent of deep-sea fishermen were employed on scallopers, 20 per cent harvested groundfish, and 3 per cent worked on herring seiners. Most vessels had a complement of sixteen to eighteen persons. Eleven per cent held the rank of captains, 32 per cent were officers (15% first mates, 11% engineers, 1% bosons, and 5% cooks) and 57 per cent worked as crew.

All deep-sea fishing companies shared an industrial model for running their enterprises, and followed many of the same managerial procedures associated with most industrialized land-based enterprises, such as modern accounting, cost-efficiency programs, and time budgeting. All vessels in the deep-sea fleet had corporate ownership; even the nominal 'owner/operator' vessels had corporate money invested in them. These vertically integrated companies with year-round operations harvested the marine resources that supplied the companies' primary, and, in some cases, secondary processing plants on shore. Driven by market demands for specific types and qualities of fish products, these companies developed and managed their enterprises to meet these needs. The demands of the market bore on the plants' production needs, and the company transmitted these requirements to the vessels where the fish were harvested. In order to assert central control, companies maintained radio contact between the vessels and their individual central office. Each day, at a specific time, the 'daily hail' took place. The company radioed the captains at sea for information on their position, catch size, and any problems they had encountered. Management also passed on additional instructions or advised captains of any change in plans. In this way vessels could be re-routed to locations where a particular type of fish could be caught; or vessels with full loads could be sent, if necessary, to plants other than their home port, where the fish would be processed.

But 'just in time' practices characteristic of assembly line plants on shore are not always compatible with sea-based harvesting.[2] Controlling vessels at sea is difficult, and controlling the harvesting of fish at sea is more difficult still. Primary production takes place at sea, where vessels chase a mobile resource in an uncontrolled environment. Uncertainty of catches – size of catch, quality and type of fish caught, timing – still prevails, and no matter how advanced the technology, catches cannot yet be accurately predicted or controlled, nor can poor weather be abated. The 'just in time' strategy for the fleet is meant to allow for a more rational use of the limited resources that the company has available to it, but it makes no concessions to the needs of the workers on board the vessels or the nature of the resource. Deep-sea fishing is actually hunting at sea. There is no guarantee that fish will be available at any particular time, or that they will remain available in any particular area. Therefore, once a crew finds fish and begins to catch them, they must continue to harvest and process the catch until there are no more fish. This work pattern, which resembles speed up and overtime in land-based facilities, results in men's doing double and even triple watches (shifts) in order to process the fish. Such 'broken watches,' because of heavy catches or breakdowns, often lead to chronic fatigue.

This 'just in time' approach used by the companies means that many vessels no longer routinely sail out of a home port. Rather, the company will route ships to specific plants where the fish will be processed most efficiently. The crews then travel by bus and/or plane to their home ports, where their wives pick them up, or the company has them driven home. These routings increase crewmen's fatigue because travel time is sometimes an additional ten to twelve hours.

The work on these vessels is also repetitive, tedious, and dangerous (Binkley 1995b; Horbulewicz 1972; Poggie 1980; Thompson 1985). The harsh and demanding working conditions of the deep-sea fishery leads to both physical and mental stress. The gruelling work schedule combines with a watch system of either six hours on and six hours off, or eight hours on and four off, punctuated by broken watches, to push men to their physical limits. Fishermen work these long hours in a dangerous and uncontrollable environment, on a continually moving vessel, with constant engine noise and noxious fumes. The actual catching of the fish takes place on an open and crowded deck in constant motion, with wires, cables, and nets, and blocks and tackles swinging overhead, the deck usually awash in the icy water of the North Atlantic. Crew often toil in wet clothing, even in below freezing weather. Most work takes place below decks under harsh fluorescent lighting in refrigerated holds or damp and slippery processing rooms. In the processing rooms men either clean or gut the fish by hand or machine. In the holds, men store the fish in ice in plastic containers or in large wooden pens.

Although most fishermen find the work repetitive, simple, and monotonous, they must pay close attention to their actions because of the high risk of injury involved in working with the machinery on board. This high level of concentration enhances fatigue. Chronic back problems, cuts and lacerations, sprains and strains, and constant mental and physical fatigue characterize the work. Even when injured or ill, fishermen continue to work their watches. While their vessels steam to and from the fishing banks they ready the gear, repair and store it, clean the vessel before landing, or rest in the cramped living quarters below deck.

Most men in this study (83%) fished for ten to fourteen days on each trip, made between eighteen and twenty-two trips per year, and had two or three days off between trips. Those men on double-crewed vessels had ten or twelve trips per year and remained ashore for approximately two weeks between voyages.[3] The remaining men took even longer trips, some up to sixty days. But no matter what their particular schedule, the cycle continued throughout the year, with additional short respites when they took off specific trips for vessel refit, vacation, deer hunting,[4] training courses, illness or accident, and Christmas. Fourteen per cent of all deep-sea fishermen in the study had been laid off sometime during that year.

Her World

Wives of deep-sea fishermen modified their work – both domestic labour and paid employment – to accommodate their husbands' fishing schedules. As the statement from Paula's mother at the beginning of the chapter indicates, deep-sea fishermen's wives had to be independent and self-reliant if they were to make a go of it. The schedule and duration of fishing trips frequently changed. When Paula's husband was fishing, she did not know when he would return, nor did she usually know where he would be fishing. By listening to the CB (Citizen's Band Radio), VHF radio, ship-to-shore radio-telephone transmissions, or by talking to wives of other crew members, she got a sense of how the fishing was going. She knew that her husband would not be coming home until he had a 'trip' (a full hold) or the fish began to spoil. She 'expects him when she sees him.' Unless he begged off a voyage for a specific event, she never planned on his being home for anything – the servicing of the car, family outings, school activities, the birth of a child, a miscarriage, or the death of a parent. Even when tragedy struck he could not come home

In the early years of marriage women have to make the transition from single person to wife and in most cases to mother. The struggle to bring up children and to manage increasing family responsibilities dominates the middle years, while the later years encompass the challenges of children leaving home, coping with ailing parents and/or grandparents, dealing with grandchildren, and facing retirement.

While all newly married women encounter many challenges associated with their new status as wives, deep-sea fishermen's wives have to confront the specific difficulties deriving from the nature of their husbands' work. Many newly married women did not realize what it would mean to have a husband who fished the deep-sea. Except for a few career women – teachers, health professionals, and businesspersons – most women (86%) gave up employment upon marriage or by the birth of their first child.[5] Then, home alone or with a small infant, cut off from their workmates and former social networks, they depended on their absent husbands for both financial and emotional support. Although most women saw this break from former friends and colleagues as part of a maturation process associated with making a commitment to a spouse and family, it created a subordinate situation for them which fostered in turn a sense of loneliness and dependency. Strategies women developed before they married, such as going out with other men while their boyfriends fished, no longer remained viable alternatives if the marriage were to last. For support and comfort they began to rely more on their extended families and to develop new friendships with women in similar situations.

Donna and Paula, both young mothers married to deep-sea fishermen, found this transition difficult. Donna and Dean had been married for three years and had a one-year-old child. Paula had been married to Peter for almost five years and had three small boys. For many newly married women, like Donna and Paula, family time stopped when their husbands sailed and recommenced when their husbands stepped ashore. Women at this stage in the life course often talked about 'rushing' their lives away, and of struggling to learn to enjoy their days without their husbands. Donna confessed to this daily struggle:

> Each day, every night, when I go to bed, it's like, well, there's another day closer to when he comes home. It's kind of sad 'cause you're rushing your life away because you can't wait until those three days come that he's going to be home. And what's three days out of three weeks, right? And then when he comes home, you want the clock to stop and then it just flies. It seems like just a couple of hours and he's back on the plane again. I think you could live and enjoy your days while they're gone, but I don't know how you do that (laughter). I haven't mastered it yet. And everyone that I know and talk to has the same general feeling that they do rush their days. Maybe now that I'm aware that I do it, I might try and not do it. I don't know how yet.

For women who did know how, breaking out of the cycle of relying on their husbands as their major emotional support represented a major step towards autonomy.

But some women found making this break more difficult than others. This chapter began with a quotation from Paula's mother, herself married to a deep-sea scalloper captain for over thirty years, in which she paraphrases a warning she gave to her daughter on her wedding day. Yet despite having grown up in a deep-sea fishing household and hearing her mother's advice, Paula still found the responsibility of maintaining the household, compounded by her separation from her husband, crippling. This struggle consumed her life:

> I just sort of walk around and do what has to be done. I deal with what needs to be dealt with at the time. I don't let it touch me emotionally. When he comes home, then I fall apart. [I] just kinda curl up in his arms and cry for two or three hours (laughter). And that I think is the worst. But you deal with it. There's nothing that can be done about it. I would never ask him to stop fishing. Never. He loves it too much. And he makes a very good living at it and takes very good care of the children and I.

As we will see in later chapters, Peter benefited from Paula's dependency on him because it ensured her subordination to him emotionally as well as financially.

As the months went by, most newly married wives developed positive ways of coping, such as forming their own support networks, throwing themselves into their own work, or starting a family. Slowly they learned to cope, first making it from one day to the next, then gradually developing a routine. As they gained control of the situation, they started developing long-term strategies that allowed them to maintain control and take daily emergencies in stride. Through this process, deep-sea fishermen's wives often achieved autonomy, personal growth, self-actualization, and independence. But for some fishermen's wives, increased independence brought a mixed response from their husbands and from themselves. For most women the household decision-making process remained a point of tension and a potential area for later conflicts. When their husbands fished, women had to make decisions which affected their households; therefore, couples had to discuss beforehand, in general terms, where husbands stood in order to develop a shared understanding about basic household issues such as finances, child-rearing, and house and car repairs. Wives tried to make the right decisions, but some times they did not. Some husbands supported their wives' decisions whether they agreed with them or not. Others did not – which tended to undermine the wife's authority and re-enforce her subordination to her husband.

But fishermen's wives are not the only group of women facing these challenges. In their study comparing commuting professional and merchant marine couples, Gerstel and Gross (1984, 164) found that 'both sailing and commuter wives also recognize a more generalized sense of independence – sometimes even suggesting that they had become "too" independent. The negative side of independence – the sense that it had gone too far – is stated much more strongly by the sailing wives than by the commuter wives. In articulating the negative side of independence, sailing wives either mean they resent how much responsibility they have to bear alone or that their independence creates problems when their husbands return.'

Fishermen's wives also struggled with this issue. For example, Donna discussed both of these negative aspects of independence:

> I'm not used to answering to anybody, because 99 per cent of the time I'm on my own, and when he's home it's nothing for me to pick up and take off, and I don't think of telling him where I'm going, and then he gets annoyed at me. It's just hard to have to answer to somebody. But if he's home more often, I might get into the habit of it. I'm just so use to picking up and going, and if he's down in the shop I don't think, 'Maybe you should tell him you're leaving' (laughter). ... It kind of gives you a little sense of satisfaction when you can successfully take care of everything (laughter). When things aren't going right, that's when I'm cursing him, like, 'Why aren't you here to do this!' But when I do it on my own, it's like, ya, I can do it. I don't need him (laughter). Then when things aren't going so well, then I want him here to fix it.

Learning to cope on her own increased Donna's self-sufficiency, independence, autonomy, and self-esteem, but it also created difficulties for her when her husband returned home. She realized that she could live alone and do the tasks her husband previously did, yet she was financially dependent on him. For some women, these psychological changes led them to question their relationship and the basis of their marriage. For some husbands, the progressive changes in their wives' behaviour led them to ask the same questions. Although the working conditions of the deep-sea fishery created the structural framework for this struggle between spouses to take place and develop, spouses seldom recognized these problems as social products of the fishery.

Being apart also jeopardized the psychological intimacy that couples might develop with daily interaction, and led instead to a psychological distancing.[6] It took time to bridge this distance, but there was very little time available between trips to build a shared history that would bring order to the lives of deep-sea fishermen and their wives. This lack of shared experience – being present at a brother's wedding, or the birth of a child, or the death of a parent – could undermine the relationship and leave a couple further apart.

Gail and Gordon had been married for over seven years and had a son and a daughter. Gail recognized that 'There's so many events that Gordon feels that he has always missed out on. Even when my son was born, I was in labour and Gordon was home with me. And that morning we got up and he had to sail that day. I was down aboard the boat and I was in labour and everything, and he sailed that night, five o'clock. I went to the hospital ten o'clock that night and I had my son the next morning at ten o'clock. Gordon did not see him till he was six weeks old.' Gail still resented the fact that Gordon could not be present for the birth of their son, and Gordon, too, felt he had missed a number of defining moments in their lives. Each individual experienced the same event in quite different ways: the husband felt left out of important events in family life and resented it later; his wife resented her husband's absence. Gail felt she had to bear all the responsibility for an event or that she could not fully participate in an event because she took part on her own – 'only one-half of a whole.' Wives frequently manipulated the timing of some events, such as a child's first birthday party, so their husbands could be present. Or they attempted to incorporate husbands into these events when possible, if only vicariously: they took photographs, or videos, or borrowed videos of events for their husbands to view on their return. Through these and other mechanisms wives tried to develop a shared history which incorporated their husbands into daily home life.

A fisherman's absence from home created other problems for the household. For children, the absence of their father could be traumatic. Young children cried out for their father when he was away. When he was home they would 'stick like glue to him.' Some children thought their father had left them because they were bad or that he didn't love them. Of course, as the children matured the nature of their demands on their fathers changed; they learned to cope with their fathers' absences, though they still did not like it. Donna wondered what it would be like for her child. She described the difficulties her friends had experienced:

> A lot of my friends have older kids. One in particular is eight years old and she's 'When's Daddy coming home?' everyday. Another friend of mine has a two-year-old and a three-year-old. Well, he [the husband's friend] took a trip off and they're used to having him home for a month, right? And every night before they go to bed, they bawl and cry and they want Daddy. I'm not looking forward to that. You want them to be close to him when he comes home, but then you almost don't because then when he leaves they miss him that much more. It's a hard situation. Same with me. If he doesn't take trips off and he's only home for three days, it's easier for me to get back in the groove of being alone.

But when the fishermen returned home a new set of challenges arose that had to be met by the household's members.

Intersecting Worlds

When their husbands came ashore most wives' daily routines changed. They began their second life – the more exciting one. The homecoming began one or two days before the vessel arrived. Wives phoned the company's hotline daily to listen to a recorded message which indicated each boat's progress and tentative return date. They cleaned the house and got all the routine chores out of the way, because once their husbands came home there would be no time for these mundane activities. Ina and Ivan had been married seventeen years and had three sons. Ina had developed a routine that allowed her to get her domestic work done prior to Ivan's coming home so she could enjoy his company: 'You do a lot more, like little house cleaning chores and all that kind of thing, when he's out. Because when he comes home, like say the day before you know that they're coming, well you race around here and you try to get everything done right up to the point. When they are at sea, you don't go nowhere. So when he comes home, well he usually wants to go here and he wants to go there. And I usually do go with him everywhere he goes. Everything is done up before he lands (laughter).' Moreover, husbands expected the home to be clean and tidied, the lawn and gardens maintained, the car serviced and washed, and their wives available. To these men, outward signs of their success reflected their status within the community. To their wives, these achievements proved their competence and capability in running their households on their own.

The boats usually docked between midnight and six in the morning, and husbands expected wives to drop everything and be waiting for them on the dock.[7] This meant bundling up young children, putting them in the back seat of the car, and driving to the wharf with a thermos of hot coffee and the sleeping children. There the wives waited for the vessel to appear on the horizon and to steam slowly to the plant's wharf. The women passed the time chatting. There was an air of excitement. And it was not just newlyweds who appeared thrilled to see their husbands. Barbara, a wife of eighteen years, described her anticipation: 'When he goes away again, in a sense you can't wait for him to come home but they're gone. It's almost like when you first go together, the minute he walks through the door, or you go to the wharf and pick him up – it's like you're first married. That's what it's like. And it's like that every trip when he comes home (laughter).'

Some husbands were more demanding than others. As one fisherman from my earlier (1986) study explained: 'When I come up over the dock, she better have the car turned around, the driver's door open, the trunk open, and be sitting on the passenger's side waiting for me. My car not being there really spites me. I just hate waiting to go home. I only have forty-eight hours' (Binkley 1995a: 58). Women's compliance with such demands also ensured that their husbands came directly home, and did not go off with their fellow crew members.

The frequency and duration of marital separation in deep-sea households continually created a tension – sexual and emotional – between couples. When a husband returned home, his spouse often had different plans for the homecoming, and disputes arose frequently from these mismatched expectations. Compromises had to be made. In any setting, separation places various forms of stress on couples and their relationship. Various demands of employment, household, and family responsibilities exacerbate these tensions between spouses. These disputes do not differ contextually from those which take place at the end of the day in households where women stay home full time and men return daily (Luxton 1980; Luxton and Rosenberg 1986; Oakley 1974; Rubin 1976). In many of these households work and family responsibilities remain separate, with work concerns dominating weekdays and social interaction limited to the weekends. Although daily interaction may also increase strains and lead to mutual allegations of taking each other for granted, it is the everyday face-to-face interaction that allows for daily tension management, emotional support, and better socialization.

For these fishermen's wives the transition from an autonomous person to a member of a couple meant relinquishing hard won independence and subordinating individual needs and desires in order to maintain their marriage. During a husband's lengthy absences a wife forged an autonomous life for herself, but she looked forward to her husband's return to give her companionship, a respite from having sole responsibility for the household, and the resumption of a social life as part of a couple. Her ideal partner was a companion, an intimate with whom she could share her life. Yet, when her husband returned home, his needs were paramount: his laundry had to be done, his favourite meals cooked, his entertainment desires met – her needs and desires were subordinated to his. Wives also recognized that during their husbands' short respites on shore, they had to rebuild their husbands – psychologically, emotionally, and physically – so that they could return renewed to the gruelling and difficult conditions at sea.

When a fisherman finally arrived home, weary and tired, he looked forward to relaxation, time with his children, and a rest from pressures. He wanted to live ten days in two: he wanted to see his buddies, participate in recreational activities, spend time with his wife and children. He expected his wife to facilitate these desires and to mediate between him and the foreign world – life on shore – that he had now entered.

A fisherman had to cope with what often became, after days or weeks away, an unfamiliar shore routine. He had to adjust to his family and leave his life at sea, the crew – his other family – and its male world, behind. He had to enter once again a matrifocal household where he might feel himself an outsider. He might find it difficult to understand why his needs did not come first, why he laboured so hard and so long to support an ungrateful family. He might feel guilty about leaving his wife and children to fend for themselves. Perhaps he feared the alienation of his children and might attempt to woo their affection with expensive toys – a four-wheel recreational vehicle for his twelve-year-old son or a boom box for his pre-teenage daughter. In some cases, a husband might begin to resent his wife and see her as the lucky one, staying ashore and enjoying family life. (This attitude was reflected in men's negative commentary about women's complaints – 'What's she griping about? She's got it so good' – common complaints among deep-sea fishermen.) As well, fishermen often found they had little in common with non-fishermen, and had no role in the community other than that of husband and father. All of these feelings might result in anxiety, insecurity, and fear of rejection.

Wives also had to balance the immediate needs of their exhausted husbands with the desire of children to welcome their fathers home. Many wives spoke of being 'torn in two.' Sometimes their children came first, sometimes their husbands. Wives' needs were subordinate to both. Yasmine, whose second husband was a coastal fisherman, felt that throughout her first marriage of almost twenty years to a deep-sea fisherman, the needs of her two children and husband always took precedence: 'because you're so wrapped up in them. I mean, okay, you're on your own basically. They go make the money and they pass it to you and that's it. You have to bring your own children up. You have to pay the bills, see that all your payments are made. You do all the cooking, the cleaning. You make up a lunch that night, and then by the time you cook supper and everything you're putting up another one, so really you have no time for yourself. It's all for your husband and your children. And you lose yourself, you don't know who you are any more. Or that's how I felt.'

Some women, like Yasmine, felt overwhelmed by the return of their husbands – yet another responsibility to take on. And for women who participated in the labour force, their husbands represented a third area of responsibility. As Margaret, manager of a retail store and married to Mark for fifteen years with one child, said: 'I always say I have three lives: my life at work, my life at home, and my life when [my husband's] home. When he's home, it's a whole different life.' For these women this 'third life' increased the pressure placed on them. Wives tried to juggle all these demands, but frequently someone felt shortchanged. Conflicts arose. For example, Gail and Gordon had a number of disagreements about what he perceived as her tendency to put the children's needs ahead of his. As Gail explained: 'That has probably created the most problems in our relationship, with regards to him feeling that I put him last, after everything else is said and done. And it's something, you know, that I've had to try to talk to him and make him realize that just because he shows up home you can't just suddenly change everything around. You can't turn around your whole lifestyle.'

Gail articulated one of the inherent problems in deep-sea fishing-dependent households: how to incorporate a husband into the daily routine of the household without losing its integrity. Some women complained that incorporating another person's needs and preferences into the household meant that routine things, like the daily cleaning and tidying, just did not get done. They recognized that it was imperative that couples take time out to talk, to do things together, and to renew their relationship, but for some it was frustrating to see the domestic chores piling up.

For others – especially full-time employed deep-sea fishermen's wives – this 'third life' offered time to relinquish responsibilities and delegate the demands and needs of the household to someone else. Gail and Gordon employed this strategy to relieve Gail of some domestic labour but also to integrate Gordon into the family's routine. As Gail explained: 'I'm the type of person that finds it much easier when he is home because he's my support person. I suppose if I had a job that wasn't as demanding. I have a career on my own ... and the demands of having a young family, and the demands of then having a husband that's gone, you combine it all together, that becomes a lot to handle. And sometimes you sit back and you wonder how you do it; why suddenly, something just doesn't just break down in the process.'

The fishermen, in response to the demands of their dual lives, often fell into two types of behaviour when they were ashore: that of 'shipmate' or 'family man.' The shipmate spent as much of his time as possible with other crew members, going home only to change his clothes, sleep, and eat. He spent little time with his immediate family. His wife might come along with him to parties or dances, but most of his activities centred upon male pursuits such as hunting and fishing, hockey and football, card playing, and drinking. Usually, the single men of the crew, and the married men whose families lived far away from the home port, dominated this group. The family man typically spent as much of his shore time as possible with his immediate and extended family, and with his close friends. He worked around the house, visited his parents, played with his kids, or went out with his wife. Seldom would he participate in activities with men associated with his life at sea, except for special functions such as a wedding reception for a crew member.

Women talked about 'taming' their husbands, that is, changing them from shipmate to family man. Through this process, husbands would gradually become integrated into the household and its routines. Gail described Gordon's transformation: 'He has gotten better than what he used to be. I'm very fortunate because, like, Gordon is the type of person, that he's very dedicated to me and the kids. I know some fishermen may come home and first thing they want to do is go out with their buddies and drink. They have to go out on their tears or their sprees. When he comes home now, he wants to be home because he wants to be with me and the kids. Like to him, that's what's important. [For a] few of his friends the number one thing is to get home and get with the boys. To him [Gordon] that shows a lack of maturity or whatever, and their responsibilities aren't right' (laughter).

The 'taming process' usually began in the first years of marriage. In some cases the process proceeded relatively quickly, in other cases it was slow and gradual, in a few cases it never took place. Gail, like many women, saw this process as a maturing one: 'a combination of having kids and recognizing the responsibility associated with the kids.' But other wives believed that it took fishermen longer to mature: 'Seems like they're in their middle thirties/late thirties when they start growing up' (Opal). Or, an alternative way of looking at this process involved a husband's recognizing that family life on shore offered greater rewards than drinking with his mates. He had a focus to his life that made his sacrifices worthwhile. Gradually a couple could forge a balance between the husband's commitments to family and crew: the husband would participate in some activities with crew members, but save time for his wife and family. Each household worked out some compromise, which was renegotiated continually through time with its changing circumstances.

Downsizing, Work Reduction, Retirement

The fisheries crisis in the 1990s led to a reorganization and restructuring of the fishing industry itself, of individual fishing enterprises, and of fishing-dependent households. In the area of Nova Scotia south of Halifax and along the shore of the Bay of Fundy, the fishery was more diverse, and the impact of the moratorium was recessionary rather than catastrophic. Deep-sea fishing companies substantially downsized their operations, tied up their vessels, sold off surplus vessels, retired aging vessels, and refitted groundfish trawlers to harvest other species such as shrimp, shark, or surf clams. The factory freezer groundfish trawlers, *Cape North* and *Cape Adare*, were sold. The deep-sea groundfish fleet (vessels over 100 feet) shrunk from thirty-seven vessels in 1986 to fifteen vessels in 1993. The deep-sea scallop fleet (vessels over ninety feet) declined from sixty-eight vessels in 1986 to forty-three vessels in 1993.[7]

Companies also substantially modified their workforce. With fewer vessels to operate and maintain, the companies needed fewer employees and less plant capacity. Some fish plants closed, while other plants had shorter production runs and/or extended vacation periods. A few 'redundant' groundfish plants were retooled to process shrimp or crab.

These newer facilities relied on technology rather than labour. Compared to labour-intensive groundfish processing, they had substantially smaller workforces, both relatively and absolutely. Moreover, most secondary processing of these products occurred elsewhere.

On vessels, companies modified their workforce by using either of two methods. The first involved maintaining one full-time crew on each of the remaining vessels and laying off all other crew. Those fishermen who retained their jobs had full employment at pre-moratorium wages. The 'redundant' workers became unemployed and thus eligible for the TAGS and/or other social assistance packages. The second involved downsizing the fleet, rotating two full crews on each of the remaining vessels, and laying off only those men who wished to take the TAGS package or retire. All of the fishermen who wanted to continue fishing could have employment, but they worked approximately half-time and at half their pre-moratorium wages. Thus deep-sea fishermen could be divided into three groups by employment status – fully employed, underemployed, or unemployed – each of which corresponded to adaptations of their respective households.

For the exceptional deep-sea households, in which the fisherman maintained his full-time job and income, cutbacks in the groundfish fishery made no direct impact on their family life. But most of the members of these households had links with other fishing-dependent households who had been affected, and all households experienced indirectly the impact of the fisheries crisis on everyday life in their communities. Moreover, many men feared that their fishing job could be affected in the next round of quota cuts.

Jane Wheelock's study (1990) in North-East England, an area of high unemployment, examined the relationship between capital restructuring and the internal dynamic of the household. Economic restructuring within the area led to a decline in employment for male manual workers, but low-paying jobs for women remained. Wheelock found that 'amongst the thirty couples interviewed there was a marked shift towards a less traditional division of household work, with men undertaking more domestic labour and childcare when they became unemployed' (1990: 3). Does Wheelock's finding hold true for this study?

Four couples – Cathy and Cory; Janet and Joe; Holly and Harry; and Elizabeth and Eric – struggled to cope with the restructuring of the fishery. Both Cory and Joe had their jobs downsized. Cory and Cathy had been married fifteen years and had three school-aged children; Cory worked two weeks on and two weeks off. Joe and Janet, married thirty years, had five adult children; Joe worked one month on and one month off. Harry's and Eric's jobs were eliminated. Harry and Holly had been married for twenty-two years and had two adult children; Harry was laid off but wanted to return to the fishery. Holly, now the primary income earner, held a full-time job as a nursing assistant. Eric and Elizabeth had three adult children and had been married for twenty-nine years. Although Eric had virtually retired, he occasionally worked as captain or mate for his previous employer. Elizabeth worked as a nurse to help support the household.

Wheelock (1990: 117) identified four types of responses – regressive change,[8] no change, some change, and substantive change – of men who were no longer working. Underemployed and unemployed fishermen responded in similar ways; however, there was a distinct difference in attitude between the laid-off fishermen like Harry and Eric who had little chance of going back to work, and fishermen like Cory and Joe who worked on doubled-crewed vessels. In the case of laid-off fishermen, men's participation in domestic labour took on two extreme forms of behaviour: no involvement with regressive or no change, and total involvement with some or substantial change. Double-crewed fishermen's participation in domestic labour was more varied. Lydia Morris's study of underemployment and unemployed workers in Hartlepool, England, may give us some additional insights into the retention of a gender-based division of labour among double-crewed and laid-off fishermen. She found that 'There are no clear-cut indications of a renegotiated division of domestic labour in response to male unemployment, however, and it seems rather to be full-time employment for women that provokes adaptation' (1995: 18).

The myth of the single male wage earner as the sole support of the household remained the ideal for most deep-sea households. The breakdown of this ideal was most dramatically seen in the laid-off fishers. Fishermen who had been laid off did not usually take advantage of their newfound 'opportunities.' Rather they felt that they had let their wives and children down by not fulfilling their side of the bargain – supporting the household. These feelings of inadequacy eroded self-esteem and self-worth.

Fishermen also frequently complained about their lack of participation in family life. Deep-sea fishermen described the three worst features of their job as health and safety hazards, high stress, and extended separation from home (Horbulewicz 1972; Poggie 1980). The most common reason for deep-sea Nova Scotia fishermen to leave the fishery – after injury and economic concerns – was the long absences from home (Binkley 1995b).

With lay-offs and the double-crewing of vessels, many husbands remained ashore for longer periods of time. This change gave couples a chance to get to know one another better. When Cathy was asked how her relationship with her husband Cory had changed now that he was home more, she replied: 'We may be a little closer than what we were when he was [gone]. When somebody is gone twelve days and home four, and twelve days and home four, that type of thing, it doesn't give you a chance to get close and talk over things that have gone on in twelve days. Like, he's gone another twelve days and you're thinking, Oh, I forgot to tell him, you know, something happened, you know, twenty days before that. Of course, then by the time he comes in, it's gone altogether.' For the underemployed or laid-off fisherman, the time ashore gave him an opportunity to rest from the pressures of being at sea, to be reintroduced to the household, and to build a shared intimacy with his wife and family. It allowed for an integration of the husband into his wife's world by relieving her of the sole responsibility of running the household and making decisions, and by opening up the opportunity for the couple to develop a common life. He also had to develop ways of coping with living ashore with increased leisure and a reduced income.

But this new-found closeness had a price for both spouses. The ideal and the reality of the husband's being home were very different. Those capacities that enabled a wife to be a supportive spouse to a fully employed deep-sea fisherman sometimes conflicted with her ability to fulfil the role of wife to a laid-off or double-crewed fisherman. For some women, having their husband home for longer periods of time could be frustrating. Many women talked about their loss of independence, of having 'another child to look after,' or that their husbands were underfoot. They were used to being on their own. If the couple had developed common interests in the earlier years of their marriage the transition appeared easier. In those cases, the wife did not feel her autonomy threatened, nor did the husband feel as though he was simply marking time until the next boat sailed.

For the fisherman with a reduced workload, being home for longer periods of time between fishing trips opened up opportunities to participate in activities with his household, kin, and community. He more readily took on tasks involving his children, especially those associated with the more pleasant aspects of child rearing – reading stories, playing – in preference to other forms of domestic labour. Some men took advantage of this time to join a sports team, or to participate in community, school, and church activities. Others took courses to enhance their fisheries credentials or to upgrade their formal education. In some cases where the husband had always participated in domestic labour in the household, he continued to do 'his chores' – the responsibilities he always took on when he was at home. Many men saw this reduction of work as an opportunity to take advantage of because they assumed that in the future they would be employed full time. Whether this assumption proved correct or not, most of these men saw the situation as a temporary one.

Unlike Cathy and Cory, who saw this time as an opportunity to be seized, Joe found it difficult to make the transition. As Joe's wife, Janet, remarked: 'I think it's hard for him to adapt to it because he's so used to fitting everything into four days that when he's home that month he's just getting used to a normal time for sleeping and a normal time for doing things when he's got to turn around and he's got to change everything again and go out.' Those under-employed men with wives in the labour force appeared more willing to do household chores, in part because they saw this situation as transient. But some men recognized that since their pay had been cut, their wives' wages now took on more importance. This was especially true if their wives earned substantial wages. In these cases, their way of thanking their wives for helping them support the family was to help with the domestic work.

But for the laid-off or retired fisherman this situation was not temporary. With no prospect of going back to sea, he had to find new meaning for his life. Holly recognized Harry's situation: 'The problem is what to do with himself – the emptiness. He mopes a lot, and that type of thing. He tries to keep busy, but deep down he's bored silly. I think his biggest problem right now is the fact that he feels that he's worthless – that he's no good to himself (laughter), that he's not productive. You try to support them, but it's still hard. He has fishing friends that are working, and then here he is, sitting home. And it's just not happened yet for him. I think he finds that really hard to deal with it. It's not easy, not at all.' No matter how well fishermen became integrated into the household, they were seldom happy spending long periods at home, and most looked forward to going back to sea or getting another job. These feelings of uselessness and boredom began to erode some fishermen's self-esteem, and as the cycle of depression deepened, fishermen might develop psychological problems.

Total involvement in the household was the other common response of laid-off fishermen. Such men needed to have a role to play and work of value. In some cases, particularly if their wives were employed, they took over their wives' work – an arrangement that required a man to have a high level of self-esteem, for in the macho culture of the fishery 'women's work' had little value. Harry had been laid off for over six months when Holly was first interviewed. Although they would never admit it, Harry and Holly had exchanged roles. Holly was working full time doing shift work at the hospital. When she took on the role of wage earner, Harry had tried to take over the household. He cooked, cleaned, and even ran her bath. Yet both found their situation unacceptable. Holly had lost her autonomy and independence; what was once an accommodation for a few days had now become a way of life. Harry had lost his self-esteem and now depended on Holly financially. Of course, Harry was 'only helping out' until he could get back to work.

With the fisheries crisis, most households experienced a decline in their household income. In most cases this meant that wives sought employment, changed their status from part-time to full-time employment, or continued their employment when they would have preferred to quit. Many women resented this change in their working conditions, not because they did not want to help out, but because they had never intended to take on this much responsibility. As Holly said, 'I always say to him, I love my work and I wouldn't do without it, but I just don't want to be obligated to having to be the one to be the breadwinner.' This sense of inadvertent betrayal permeated both men's and women's discussion of changes wrought by the restructuring of the fishery.

Coping with retirement resembled the challenges faced by households of laid-off fishermen. Some fishermen saw retirement as a well-earned reward for a lifetime of hard work, while others saw it as being deprived of what they loved to do. Whether retirement was joyously anticipated or not, once the fisherman was retired or semi-retired, he had to make adjustments to this new way of life, which affected both spouses. Before retirement, couples had developed strategies for living together, predicated on independence and autonomy. The wife had gone from having children at home, to having no one at home, to having her spouse home part- or full-time. In some cases, couples realized that although they had lived together for many years and raised a family, they had no common interests except for their children and grandchildren. For twenty to twenty-five years his life had revolved around fishing, hers around the family and the community. In all cases, couples had to forge new strategies to build a different kind of life together, or decide to go their separate ways.

The deep-sea fisherman's approach to retirement greatly depended on whether he had developed interests outside of the fishery or not. For some, retirement was an opportunity for the leisure they always wanted – to play golf, putter around the garden, collect stamps, or take classes. For those who had no leisure interests, the prospect of retirement was agonizing, and they continued to fish until they were forced out. Being home full time created an additional strain on the relationships of many couples. Elizabeth, whose husband, Eric, had been at home for eight months in semi-retirement – subbing for mates and captains on other vessels – explained concerning their adjustment:

> Eight months is a long time to live with him constantly. There's only been probably – in thirty years of marriage – several times that he's been home for five, six, or seven months at a time for one reason or another. So it was rather unusual for me to have him around that much, and I do find that different. On the other hand I try to understand how difficult this is for him. I mean he was used to having a job that was at least, probably nine or ten months of the year, and suddenly he's only working sort of part-time. Often when, even years ago, it seemed like you had to be responsible for entertaining him, so to speak, because when he was home he had certain things that he did, and he came and went. Like when he was home, he didn't have a lot of things that he did.

Nothing had prepared Eric for retirement. He had not developed any interests outside the fishery. He had counted on Elizabeth and his fishing buddies to mediate the shore world for him. He kept hoping to go back full time, but he knew that he could not physically or mentally handle the pressure and stress of full-time employment. So he was developing a daily routine while he waited in vain. Before Eric retired, Elizabeth had encouraged him to explore alternative interests, such as curling and golf, which would sustain him when he left the fishery. As Elizabeth said:

> I wish he would have a few more things that he did on a regular basis. His day is so routine and not very satisfying, I'm sure, to him. He gets up, goes down to the local minimart for coffee, and visits around [with] some of the other men that are in a similar situation, and [he] spends a fair amount of time watching TV, which is not very satisfying for him. He does golf a little,

and he does curl a little, but he just doesn't go into it enough in the winter. I think he thought, well, if he really got involved, well then if he got a job then what would he do? But I say, don't worry about that if something major comes along and you can't do the curling, it's only your dues. It's no big deal.

Throughout this chapter, we have seen how men's and women's work worlds were separate. While men's work generated income for the household, women's work was crucial to the day-to-day operation of the household and to supporting the male wage earner emotionally, psychologically, and physically so that he could continue to work in the fishing industry. With the downsizing of the fishery, their worlds collided. For all three types of fishermen's households – employed, underemployed, and unemployed – it was assumed that women would continue to support the fishing industry by sustaining their husbands' well-being and by helping them cope psychologically with their new working conditions. Women were also expected to take on paid employment to augment their households' reduced income in the case of double-crewed fishermen, or to replace the income of laid-off or retired fishermen. In the next chapter, we will explore in greater detail how domestic labour supports fishing-dependent households.

CHAPTER FOUR

Running the Household

I do most of the household stuff. He helps out. (Hazel)

Both wage and domestic labour were needed for the survival of the fishing-dependent household. Livelihood strategies employed by members of these households, whether based in the coastal or the deep-sea fisheries, defined and constrained domestic labour.[1] In many cases the demands of the fishing industries and women's paid labour outside the home conflicted with the demands of domestic labour. Husbands and wives in fishing-dependent households, like all households, tried to balance these demands in such a way as to meet their households' domestic and financial needs while maintaining their own individual autonomy. Through domestic labour each member of the household contributed to its daily and long-term survival. These practices produced household dynamics stretching over a lifetime. Women, by virtue of their role as wife and mother and by their husbands' frequent absences, had primary responsibility for the household and childcare, while men 'helped out.' Although some husbands participated more than others in these tasks, their contributions had to be constantly renegotiated throughout the life cycle, the fishing cycle, and the changing employment patterns for both spouses. But wives found these negotiations difficult, and in general husbands resisted increasing their level of domestic involvement (McMahon 1999). The challenge lay in 'defeminizing' domestic labour and thus changing the patterns of men's participation in it.

Dividing Up Household Tasks

Now, let us focus on how couples in these fishing-dependent households contributed to the reproduction of their households through unpaid labour. We will look at a range of activities associated with domestic labour in order to determine in a rough fashion what contributions these men and women made to their households. The intention of this discussion is not to arrive at a complete picture of all the work done by particular couples, or to estimate the total number of hours each spent engaged in domestic labour. Rather, the aim is to get some picture of the extent to which gender segregation existed in the performance of domestic tasks and to consider the circumstances in which such segregation was relaxed or abandoned. These patterns should be seen as indicative rather than fully representing domestic work.

I chose nine domestic chores – cleaning, washing clothes,[2] cooking, putting out the garbage, grocery shopping, mowing the lawn, snow shovelling, and tidying – as rough indicators of the gender-based nature of domestic work. Cleaning, washing clothes, cooking, grocery shopping, and tidying represent conventional female tasks. The remaining chores – putting out the garbage, mowing the lawn, and snow shovelling – represent conventional male tasks.[3] I asked women how frequently they did these domestic chores – never, less than once a week, once a week or more – when their husbands were at sea. I then asked how frequently they did the same tasks when their husbands were home. Tables 4.1 and 4.2 summarize these results for the coastal and deep-sea samples, respectively.

Analysis of these data indicated that although individual households had their own personal preferences concerning what chores individuals did, with whom, and when, domestic labour was fundamentally constrained by the need to make a living. For example, over 90 per cent of all women cooked, cleaned, and tidied more than once a week. In coastal fishing-dependent households the frequency of wives doing these tasks did not vary substantially throughout the year. Coastal fishermen usually returned home every night or stayed away for only short periods of time during specific fishing seasons. In fact, many coastal fishermen remained at home continuously during the winter months. Given these practices, women's domestic work in these areas varied little in relation to either the presence or absence of their husbands, or to the seasonal fishing cycle. However, in deep-sea households, women's domestic work patterns changed dramatically depending on whether their husbands were at home or away. Tasks took on different priorities. For example, according to our survey, when deep-sea husbands returned from sea their wives cleaned the house less (almost 43 per cent changed from cleaning once or more a week, to less than once a week),

Table 4.1 Frequency of Domestic Chores Done by Women for Coastal Sample (in percentages)

Domestic chore	Husband away			Husband home		
	Never	Infrequent[1]	Often[2]	Never	Infrequent[1]	Often[2]
Cleaning	1.7	1.4	96.9	1.3	2.1	96.7
Clothes washing	0.7	0.7	98.6	0.7	0	99.3
Cooking	3.4	1.3	95.3	2.0	0.7	97.4
Garbage	18.1	13.4	68.5	56.0	11.3	32.6
Grocery shopping	2.0	23.6	74.5	3.4	18.6	78.0
Lawn mowing	45.0	24.9	30.2	56.7	19.3	24.0
Snow shovelling	49.0	20.9	30.2	63.3	16.6	20.0
Tidying	0	0.7	99.3	0	0.7	99.3

Missing values (2)
[1] Infrequent = less than once a week
[2] Often = once a week or more

Table 4.2 Frequency of Domestic Chores Done by Women for Deep-Sea Sample (in percentages)

Domestic chore	Husband away			Husband home		
	Never	Infrequent[1]	Often[2]	Never	Infrequent[1]	Often[2]
Cleaning*	1.3	4.7	94.0	12.0	42.7	45.3
Clothes washing*	2.0	20.7	77.3	2.7	0.7	96.6
Cooking**	4.7	1.4	93.9	4.7	1.4	93.9
Garbage*	8.7	26.7	64.7	51.3	17.3	31.4
Grocery shopping**	0.7	17.5	81.9	6.0	10.1	83.9
Lawn mowing*	38.0	22.0	40.0	68.0	15.3	16.7
Snow shovelling**	33.6	28.8	37.6	68.0	14.0	18.0
Tidying***	1.4	0	98.6	2.7	0.7	96.7

Missing values *(0); **(1); ***(2)
[1] Infrequent = less than once a week
[2] Often = once a week or more

but they washed clothes more often (a little more than 19 per cent increased their wash to once or more a week). Why these changes?

As mentioned in the previous chapter, deep-sea fishermen and their wives liked to spend his shore leave engaged in leisure or family activities. Only the essential tasks were done, such as washing his fishing clothes. Men returned from the sea with at least two weeks of dirty clothes, which women washed and dried before their husbands' departure. Normally this laundry had to be processed in two or three days. Women usually cleaned their homes thoroughly before their husbands' return and did only 'necessary' cleaning and tidying when he was home.

Women also changed their eating patterns according to their husbands' work schedule. Some women never cooked when their husbands were away and then cooked almost all the time when they were home; for other women the reverse was true. A few husbands came home and took over the kitchen (although they infrequently cleaned up afterwards), and a number of them wanted to barbecue every evening. But whether eating in or out, the husbands' food preferences prevailed. Deep-sea fishermen's wives commonly complained about the amount of weight they gained when their husbands came ashore.

Women have conventionally done the grocery shopping, but coastal fishermen's wives spent less time grocery shopping than their deep-sea counterparts. Why? Coastal fishing-dependent households tended to live in smaller communities, and they frequently bought their groceries in the larger centres. These households seldom made a 'food run' more than once a week. Deep-sea fishing-dependent households tended to live in the bigger centres, had more access to grocery stores, and shopped more frequently than their coastal counterparts. About 3 per cent of coastal fishermen's wives and 6 per cent of deep-sea fishermen's wives never did the grocery shopping when their husbands were home and only infrequently shopped for groceries when he was away. In households where men managed the money and controlled the financial resources, they also took on the primary responsibility for banking and for purchasing commodities including groceries. (Later, in chapter 8, which examines household finances, we will see that some husbands (e.g., Peter) used the purchase of commodities as a part of their strategy to control their households, their finances, and their wives, when they were away.)

In fishing-dependent communities, men most frequently engaged in domestic labour by doing conventional male chores – putting out the garbage, mowing the lawn, and shovelling the snow. In comparison to their coastal counterparts, deep-sea fishermen's wives more commonly performed these tasks and did them more frequently (see Tables 4.1 and 4.2). In coastal households where men seldom went fishing for long periods of time, women could put off these chores until their husbands came ashore. If necessary, women would do the chores themselves or have another male family member do them. However, in most of the deep-sea fishing-dependent households, where the wife never performed these chores, a son, another male relative, or 'a boy from the neighbourhood' did the task when the husband could not. In some cases, in the deep-sea households, wives hired someone to cut the lawn or shovel the snow. But in most cases the wives handled these tasks.

Many fishermen also looked after home repairs and maintenance. Some men were skilled at carpentry, plumbing, and electrical work, which allowed them to do their own home renovations and help other family members with such domestic projects. However, more typically, the wives of deep-sea fishermen did such maintenance, repairs, and renovations themselves. Of course in an emergency, or when they could not do this work themselves, the wives would ask for help from friends or family, or they would hire someone.

Maintaining the household also involved using community-based services – banking, looking after car maintenance, trips to the drugstore, and using legal, medical, and dental services. Tables 4.3 and 4.4 summarize the frequency of using of these services. The use of legal, medical, and dental services followed similar patterns for both groups. About 10 per cent of women never went to the dentist. About 70 to 75 per cent of women never went to a lawyer. One to 2 per cent of women never went to a physician. About 3 per cent of coastal fishermen's wives, compared to 10 per cent of deep-sea fishermen's wives, never did the banking.

Most women saw taking the car to the garage as a 'male thing.' About 30 per cent of women in deep-sea households and about 38 per cent of women in coastal households had never taken the car to the garage. In coastal households, some women spoke of 'helping' their husbands out by taking the car in to be fixed. Others said that it was part of their job as shore skippers to keep the family vehicle running. Women married to deep-sea fishers did see it as primarily 'his job,' but if the car broke down when their husbands were away, if it had to be done they did it.

In rural Atlantic Canada, going to the post office entailed more than just picking up parcels and buying a few stamps. In many small towns, home delivery did not exist. The post office acted as a community centre of sorts. People developed routines around when they picked up

70 Set Adrift

Table 4.3 Frequency of Use by Women of Community Services for Coastal Sample (in percentages)

Service	Husband away			Husband home		
	Never	Infrequent[1]	Often[2]	Never	Infrequent[1]	Often[2]
Bank	3.4	52.3	44.3	3.3	50.7	46.0
Garage	37.6	61.0	1.3	54.7	44.6	0.7
Dental services	10.7	88.0	1.3	12.0	86.6	1.4
Drug store	1.3	51.1	47.6	0	49.3	50.7
Legal services	73.8	24.2	2.0	75.3	23.3	1.4
Medical services	1.3	98.0	0.7	0	99.3	0.7
Post office	13.4	53.0	33.6	16.7	49.7	33.6

Missing values (0)
[1]Infrequent = less than once a week
[2]Often = once a week or more

Table 4.4 Frequency of Use by Women of Community Services for Deep-Sea Sample (in percentages)

Service	Husband away			Husband home		
	Never	Infrequent[1]	Often[2]	Never	Infrequent[1]	Often[2]
Bank*	10.0	50.7	39.3	18.7	42.6	38.7
Garage**	30.2	69.8	0	63.1	36.2	0.7
Dental services***	9.5	89.8	0.7	12.9	86.4	0.7
Drug store***	1.4	54.6	44.0	6.8	51.0	42.2
Legal services**	70.5	28.2	1.4	74.3	24.3	1.4
Medical services***	2.0	95.2	2.7	4.1	92.5	3.4
Post office****	18.5	35.7	45.8	22.6	35.6	41.8

Missing values *(0); **(1); ***(2); ****(4)
[1]Infrequent = less than once a week
[2]Often = once a week or more

their mail. Everyone in the community knew when the mail should be sorted. Some timed their pickup to coincide with this event. Others picked up their mail when they came home from work or after they dropped off their children at school. Some folks just hung out at the post office to talk with their neighbours. About a third of coastal fishermen's wives and about 45 per cent of deep-sea fishermen's wives routinely picked up their mail. In deep-sea households, husbands frequently went to get their households' mail. It was a good place for them to catch up on the local gossip and to 'see folks.' But their wives also saw it as an advantage. As one woman explained: 'You can count on him being gone for most of the morning so I can get my work done.'

In general, in coastal fishing-dependent households domestic chores segregated along conventional gender lines (Frederick 1995; Jackson 1992). This segregation broke down in deep-sea fishing-dependent households because the men were often away too long to be counted on to do the conventional 'male chores.' For many deep-sea households this pattern continued when the husband returned home. In these households we saw a feminization of conventional male household jobs.[4]

I am not saying that men did not do any conventional female chores; rather they did these tasks on what was seen as a voluntary basis – as doing something out of the ordinary, something special. When I asked fishermen's wives if their husbands did any conventional female household chores, most women responded as Hazel did at the beginning of this chapter: 'He helps out.' By this, they meant that their husbands did specific tasks when asked, or in some cases when *persistently* asked. Men made it clear that they viewed the house and children as their wives' responsibility, and that they gave their labour voluntarily. When I asked women what their husbands did specifically to help, they replied; 'Occasionally he does the dishes,' or 'Once or twice a month he cooks:' or 'He vacuums, if he's asked.' These three chores – doing the dishes, vacuuming, and cooking – plus washing the floors, comprised the most common household tasks done by men. But it should be noted that doing dishes frequently meant putting dirty dishes in the dishwasher, or emptying the dishwasher of clean dishes, or drying the dishes while wives washed them. Cooking, in some cases, meant turning on the oven, preheating previously prepared dishes, reheating leftovers, or barbecuing meat to go with the rest of the meal cooked by the women. Only a few men cooked a full meal 'from scratch.'

Instead of relying on their husbands to help out, women frequently relied on other family members, usually female kin – mothers and daughters – living in the household to do domestic chores on a regular basis. These women were not asked to do these tasks, rather they had been delegated (implicitly or explicitly) these chores. For example, young girls did the dishes while older children, usually girls, prepared simple meals and helped with laundry or vacuuming. Most children cleaned their rooms and made their beds. In some deep-sea fishing-dependent households both kin and non-kin received pay to do domestic labour. This practice most often prevailed when the wife had a well-paying job. In some cases an unemployed adult daughter was hired to clean the house. In other cases, a cleaning person would come in on a weekly or biweekly basis. Coastal fishing-dependent households, on the other hand, seldom employed domestic help. Thus in the majority of these fishing-dependent households domestic labour remained feminized, as did child-rearing.

Looking After Children

Each stage in the life cycle puts particular stresses on a household. Household relationships, like all social relationships, involve power. As partnerships form or end and as household membership varies over the life course, so power relationships within the household change. The accommodations that couples make during the first years of marriage – single persons transforming themselves into wives and husbands and in many cases into mothers and fathers – differ from those accommodations made when the couple struggles to bring up children and to manage their increasing family responsibilities (Fox 1997). The challenges associated with the later years of marriage – children leaving home, husbands retiring, and looking after ageing parents and grandchildren – can also modify power relations. Changes in work patterns of household members also disrupt these power relations and lead to new accommodations (Brayfield 1995).

Tables 4.5 and 4.6 summarize the demographic characteristics of families represented by the households under study. On average, women had two children; a little less than 7 per cent of women had no children. Family size did not vary greatly; about 81 per cent of women had families with one to three children (see Table 4.5). The age distribution of children within families was also similar. For coastal fishing-dependent families approximately 15 per cent of children were pre-schoolers, 43 per cent were school age (elementary, junior high school, and high

Table 4.5 Distribution of Number of Children per Family (in percentages)

	Coastal sample	Deep-sea sample
Mean number of children	2.1	2.2
Number of children/families		
0	6.6	6.7
1	23.2	22.7
2	38.4	35.3
3	19.9	24.0
4	7.9	5.3
5	2.6	4.0
6 or more	1.4	2.0

Missing Values (0)

Table 4.6 Age Profile of Children by Type of Fishing Enterprise (by frequency and percentage)

	Coastal		Deep-sea	
Age range	Total	%	Total	%
0–4	48	14.9	42	12.9
5–9	49	15.2	58	17.8
10–14	49	15.2	46	14.2
15–19	41	12.7	48	14.8
20–24	34	10.5	53	16.3
25 and over	102	31.6	78	24.0
Total	323	100.1	325	100.0

Missing values (0)

school students), and 42 per cent were adults (20 years and older). For deep-sea fishing-dependent families approximately 13 per cent of children were pre-schoolers, 47 per cent were school age, and 40 per cent were adults. Since the median age for both groups of women in the study (38/39 years) and their length of marriage (14/15 years) were virtually the same, the differences in profile of children's ages can be attributed to stochastic variation (see Table 4.6).

Women's paid employment created additional stresses on house-

Table 4.7 Age Profile of Children by Type of Fishing Enterprise Where Mother Was Employed (by frequency and percentage)

Age range	Coastal		Deep-sea	
	Total	%	Total	%
0–4	17	11.5	15	11.1
5–9	10	6.8	20	14.8
10–14	19	12.8	17	12.6
15–19	28	18.9	31	23.0
20–24	21	14.2	27	20.0
25 and over	53	35.8	25	18.5
Total	148	100.0	135	100.0

Missing values (0)

holds. At the time of their interviews approximately 47 per cent of all women had jobs. The age distribution of children in families with 'working mothers' indicated different patterns for the two samples (see Table 4.7). Deep-sea fishermen's wives (re)entered paid employment earlier in their marriage than coastal fishermen's wives did. They also had proportionately more dependent children than their coastal counterparts.

As the composition of the household and livelihood strategies changed throughout the life course, demands on domestic labour varied as well. Table 4.8 indicates the primary childcare givers for pre-school age children in both samples. In deep-sea fishing-dependent households, the more pre-school age children in the home the greater the probability that the mother would be the primary childcare giver. When there was one pre-school-age child, then 79 per cent of mothers were primary childcare givers. When there were two pre-schoolers, then 91 per cent of mothers were primary childcare givers. When there were three pre-schoolers, then all mothers were the primary childcare givers. Although most of these women had a job when they married, many gave up their employment after marriage or at the birth of their first child. The above pattern reflected an inverse relationship between workforce participation patterns and childbearing: the more pre-school-age children one had the less likely one would be employed.

Those women who continued to have paid employment had to rely on other people as primary care workers. In the deep-sea sample, where there was only one pre-school-age child, about 6 per cent of women had

Table 4.8 Primary Childcare Giver for Pre-school-age Children (in percentages)

Type of minder	Coastal sample*		Deep-sea sample**		
	First child	Second child	First child	Second child	Third child
Self	78.9	66.7	79.4	91.7	100
Relative	2.6	0	5.9	8.3	0
Unpaid	7.9	16.7	2.9	0	0
Paid	10.5	16.7	5.9	0	0
Daycare / nursery school	0	0	5.9	0	0
Number of households	38	6	34	12	2

Missing values *(10); **(8)

a relative as the primary childcare giver, about 6 per cent employed a paid worker in the home, about 6 per cent used a formal daycare or nursery school; and about 3 per cent had an informal co-operative arrangement with a friend or neighbour. With the birth of the second child, most women relied on their mothers or their husbands' mothers, and possibly their husbands, when they were home, to look after their children. When the third child was born, the wife gave up work and stayed home. By having a relative, particularly a mother or mother-in-law, look after their children women paid less than they would in the formal childcare sector. In every situation the wife's childcare responsibilities, the family's ability to pay for daycare, or the ability to put up with the difficulties inherent in depending on informal daycare, limited the wife's ability to be employed.

The coastal sample however, indicated another pattern. When there was only one pre-school-age child, almost 79 per cent of women (approximately the same percentage as in the deep-sea sample) were the primary childcare givers. However, when there were two pre-schoolers in the household, only two-thirds of the women were primary childcare givers. These results seem counter-intuitive. However, we must remember that most wives in coastal fishing-dependent households participated, directly or indirectly, in the household fishing enterprise. For some activities, such as baiting trawl or running errands, women found that one pre-school-age child only slowed down their work but two pre-schoolers made it impossible to work. Therefore, women with more than one child needed childcare in order to continue their tasks related to the household fishing enterprise.

Coastal fishing-dependent households relied more on co-operative and paid help than on relatives or daycare or nursery school. Less that 3 per cent of women with one pre-school-age child depended on their mothers or mothers-in-law to help with their children. None of the women with two pre-schoolers relied on their relatives. In coastal households, even when the wife's mother or mother-in-law agreed to take care of the first child, there was reluctance – on both their parts – to continue this arrangement after the birth of a second child. Wives believed that the level of social control exerted by their mothers and mothers-in-law was not worth the convenience of having these women as primary childcare givers.

The nature of women's work in the coastal fishery varied frequently and required a great deal of flexibility in their routines. This meant that the nature of childcare had also to be flexible. Formal daycare and nursery schools catered to children who needed daycare throughout the day, every day. Moreover, they were usually located in larger centres along the shore, not in the rural communities where most coastal fishing-dependent families lived. Short-term, flexible 'babysitting,' whether paid or cooperative in nature, made the most sense for these women and about an equal number took advantage of these arrangements. Thus about 11 per cent of women with one pre-school-age child and about 17 per cent of women with two pre-schoolers relied on paid help of this nature. About 8 per cent of women with one pre-school-age child and about 17 per cent of women with two pre-schoolers relied on cooperative arrangements. These informal exchanges of labour also facilitated social activities such as going to the movies or to bingo, which might not have been possible if these women relied on relatives.

However, both groups of women did rely on relatives and friends to help out for short periods of time. Short-term babysitting – for an hour or so – took place on a more informal level. Here sisters, grandparents, aunts, neighbours, and older children, usually daughters, could be called upon to help out on short notice. Deep-sea fishermen's wives relied on these informal arrangements more heavily than their coastal counterparts.

But women who did not have kin living in the area found the lack of informal childcare and family support difficult. As Joan, a participant in my earlier study in 1987, who was originally from Newfoundland, noted: 'I don't know how we coped that first year and a half. Plus we had no car

Table 4.9 Type of Minders for School-age Children by Sample (in percentages)

Type of minder	Coastal sample	Deep-sea sample
Relative	35.9	50.0
Unpaid	5.1	28.1
Paid	17.9	46.9
After school program or daycare	17.9	40.6
Number of households	39	32

and the baby on top of it ... At three o'clock in the morning, Amy'd be crying and I'd be crying, because I just had no one. My mom was in North Sydney. I was the oldest of nine, seven of us living. Mom still had children at home. Mom couldn't drop everything and come. These women here had that advantage' (Binkley 1994: 89). Women who lived in isolated coastal fishing communities had to develop support networks among their neighbours and possibly among their husband's kin. Women who lived in larger centres had access to formal daycare centres, but most still relied on friends in the community. No matter how much emotional support was available, these women shouldered the responsibility of raising their children alone. They depended much more heavily on cooperative arrangements with other women than did women with relatives living nearby.

Caring for older children created a different set of challenges. Whether pre-school-age children or school-age children, most fishermen's wives continued to be the primary childcare providers, but they relied heavily on kin living in the area for support, especially for informal, short-term, and unexpected childcare. Table 4.9 summarizes the types of help both coastal and deep-sea fishing-dependent households use. Deep-sea households relied much more heavily than coastal households did on help from outside the home, particularly for after-school activities and at night. Their reliance on outside help reflected their husbands' absences and their relative affluence. Those women who had no kin in the area or had a job were the hardest pressed.

Women were the primary childcare workers, but how involved were the husbands in raising their children? Tables 4.10 and 4.11 summarize husbands' involvement with their pre-school-age and school-age children. In general deep-sea fishermen indicated two patterns on the

Table 4.10 Husbands' Interaction with Pre-school-age Children (in percentages)

Activity	Coastal sample (N = 48)*			Deep-sea sample (N = 36)**		
	Never	Infrequent[1]	Often[2]	Never	Infrequent[1]	Often[2]
Feed	27.1	35.3	37.6	19.4	13.9	66.7
Diaper	70.8	16.7	12.5	52.8	8.3	38.9
Dress	41.7	20.8	37.5	27.8	8.3	63.9
Bathe	50.0	22.8	27.2	55.6	13.9	30.5
Play	4.2	12.5	83.3	5.6	11.2	83.2
Get up	45.8	27.1	27.1	55.6	16.7	27.7
Put down	25.0	18.8	56.2	22.2	22.3	55.5
Get up at night	66.7	14.5	18.8	52.8	27.8	19.4
Take to sitters	91.6	2.1	6.3	66.7	13.9	19.4

Missing values *(0); **(6)
[1]Infrequent = less than once a day
[2]Often = once a day or more

extremes: total involvement or no involvement with their children. Most deep-sea fishermen, such as Donna's husband, Dean, took no responsibility for their children. When left to babysit, Dean frequently bundled up the child and went to his mother's or mother-in-law's home so that she could look after the baby. Dean felt that it was women's work to take care of children. A few husbands, such as Gail's husband Gordon, would come home and take full responsibility for the children, giving their wives a good break.

Men liked to participate in children's activities that were the most enjoyable for them.[5] Whether coastal or deep-sea fishermen, most men (83%) played with their children at least once a day. Although over 50 per cent of husbands never bathed their children, over a quarter did this task at least once a day. Over 55 per cent of men daily tucked their children into bed, although almost a quarter never did this.

In general, fewer coastal fishermen engaged in these tasks than did their deep-sea counterparts. However, those coastal fishermen who did engage in these chores, did so more frequently than did deep-sea fishermen.[5] Approximately 71 per cent of coastal fishermen never changed a diaper compared to fifty-three per cent of deep-sea fishermen. Moreover, 39 per cent of deep-sea fishermen changed diapers at least once a day compared to 13 per cent of coastal fishermen. About two-thirds of

Table 4.11 Comparison of Husbands' and Wives' Interaction with School-age Children (in percentages)

Activity	Coastal sample (N = 74)			Deep-sea sample (N = 78)		
	Never	Infrequent[1]	Often[2]	Never	Infrequent[1]	Often[2]
Husbands' interactions						
Getting children ready for school	47.3	28.4	24.3	50.0	32.1	17.9
Making a lunch to take to school	97.3	1.4	1.4	87.2	9.0	3.8
Looking after children after school	21.6	37.9	40.5	19.7	27.7	52.6
Supervising homework	36.5	44.6	18.9	31.2	46.7	22.1
Wives' interactions						
Getting children ready for school	14.9	8.1	77.0	10.3	3.8	85.9
Making a lunch to take to school	95.9	1.4	2.7	80.0	5.3	14.7
Looking after children after school	14.9	10.8	74.3	11.7	10.4	77.9
Supervising homework	7.8	23.4	68.8	7.8	23.4	68.8

[1] Infrequent = less than once a day
[2] Often = once a day or more

coastal fishermen did not get up in the night while a little more than half of deep-sea fishermen remained in bed. About the same percentage of fishermen reported getting up at least once a night. Forty-six per cent of coastal fishermen never woke up their children compared to 56 per cent of deep-sea fishermen. Again about the same percentage got their children up at least once during the day. About 42 per cent of coastal fishermen never dressed their children; 28 per cent of deep-sea fishermen never dressed theirs. Moreover two-thirds of deep-sea fishermen dressed their children at least once a day compared to one-third of coastal fishers. Two-thirds of deep-sea fishermen and over 90 per cent of coastal fishermen have never taken their children to the sitter's, while about a fifth of deep-sea fishermen and 6 per cent of coastal fishermen regularly did so.

These same trends continued with school-age children, which involved tasks such as getting children ready for school, making their lunches, looking after them after school, and supervising their homework. As expected, women interacted more with their children. Coastal households showed a more varied response, while in deep-sea fishing-dependent households, husbands' involvement with childcare still appeared to be a continuum, with overrepresentation at the two extremes: total involvement or non-involvement.

It also appeared that in deep-sea households, husbands' participation in domestic labour related to the amount of time onshore between fishing trips. The longer the time onshore the greater the likelihood that husbands would become involved in domestic labour. Men expected and women accepted that a deep-sea fisherman needed at least two or three days to recuperate from his work at sea. As Barbara said of her husband: 'after a few days ... when he gets rested, he'll help, and he knows that I'm really busy or that I have a hundred and one things going on here, so he does, he chips in.' There seemed to be an understanding in many deep-sea households that if a husband's layover between trips was greater than two or three days, then most wives expected some relief from their household responsibilities at the basic level of 'helping out,' such as drying the dishes, doing the laundry, or vacuuming. Many wives considered any work from their husbands during the first two or three days of shore leave a bonus.

Helping Out around the House

Other researchers, notably Davis (1983b), Porter (1983), and Sinclair and Felt (1992), have explored the changing patterns in the division of labour in Newfoundland fishing-dependent communities.[6] This study builds on those findings and indicates more nuanced patterns of the division of labour between and among coastal and deep-sea fishing-dependent households. As intimated earlier, in these households there remained an underlying belief that men should be the primary financial support of the household and that women should have the primary responsibility for the household and children; that both spouses should try to help each other with these responsibilities, but that this help was voluntary. In the case of women, they relied on their children and kin, usually female, to help. Men generally helped out when asked, and in case of extraordinary events, but their help was limited to these conditions. Let's see how these expectations played out in different types of fishing-dependent households, beginning with deep-sea households.

In deep-sea households with a single male wage earner, the division of labour remained the most rigidly defined along gender lines: men and women saw their spheres of responsibility as completely separate. Here the husband took on the full responsibility of earning money to support the household, which rendered his wife economically dependent on him. The wife, who was in a subordinate role, had to assert her control over the household in order to redress her power imbalance. Most wives in this situation asserted that their husbands should not have to participate in household chores at all. As Janet, who had been married to Joe for thirty years and raised five children, stated: 'Well, if you were away working sixteen hours a day for three weeks, and home for four, would you be expected to come home and do the housework?' Janet saw her domestic labour as equal to her husband's financial contribution to the household. To have her husband engage in domestic labour in any way undermined her sense of self-esteem and her base of power. However, Joe repaired, renovated, and maintained the physical structure of their home and did other 'manly' chores that Janet suggested to him. In most households where the husband was the sole wage earner, his non-involvement in the domestic labour was perceived as an earned right and as part of his identity as a successful fisherman. Similarly, women in these households measured their success by the degree to which they ran the household on their own without their husband's help.

However, one household with a single male wage earner seemed to have one of the highest levels of integration of all households in the study. Paula had been married to Peter for almost five years and had three small boys. When Peter, a deep-sea fisherman, returned from the sea, he took over much of the care of their three young sons, all under five, thus giving his wife 'a break.' But Paula was not free of household responsibilities; rather, she was expected to cook all of Peter's favourite meals and to entertain his friends and family. By taking care of the children, Peter freed up Paula, enabling her to cater to all of his other needs. But the power relations in this household reflected the breakdown of Paula's control over her sphere of influence. Paula did not control her domestic labour, which was not highly valued by Peter. She had no base for power within the household. Paula also had no access to the household finances, except for a small allowance Peter gave her. Peter controlled the household domain as well as his own, thus ensuring Paula's subordination to him.

But what happened in dual income fishing-dependent households, where wives had full-time professional jobs?[7] The most exceptional example from our study – Gail and Gordon – creates a counterpoint to Peter and Paula. Gail and Gordon had been married for seven years and had a son and a daughter. Gail had full-time professional work, and Gordon, a deep-sea fisherman, fished for three months and was home for three months. During the months he remained home, Gordon had primary responsibility for childcare and the household chores – laundry, cooking, cleaning, household repairs, mowing the lawn and so on. When he went back to sea, his mother took over responsibility for the children, and Gail did the other household tasks or hired someone to do them for her. Throughout the cycle, whether Gordon was home or not, Gail remained in charge of the domestic domain. Both Gordon and his mother deferred to Gail on household matters. This was not a situation of joint control where the couple discussed what and how things should be done. Gail made the decisions and the others helped her out.

Other deep-sea couples with a dual income had a less integrated approach to domestic labour than did Gordon and Gail, but in general they had a more integrated approach than households with a single male wage earner. These dual income professional households also depended most often on hired help to clean the house, look after the children, do snow shovelling, lawn mowing, and other yard work, and to do routine repairs and maintenance.

But Gail and Gordon's arrangements were the exception. Gordon's involvement in household tasks came from the combination of his availability to do the work because of his long shore leaves and Gail's need for help because of her full-time employment. Although deep-sea fishermen with wives employed in the labour force more frequently volunteered to do household chores on an ad hoc basis, they did not take on these chores as part of their 'work.' Men saw themselves as just helping out and made it clear that the household remained their wives' responsibility. Many fishermen with employed wives saw their wives' working as helping them support the family, so the men repaid them by helping with household chores. Once a wife stopped working, her husband stopped helping out.

In coastal fishing-dependent households the division of labour appeared to be firmly defined along gender lines but was also mediated through this practice of 'helping out.'[8] In these households, childcare and domestic labour became part of the exchange of services husbands and wives negotiated daily between the household and the fishing enterprise. Take, for example, the most 'unconventional' household in the study – Hazel and Henry, who had been married for twenty years and had a daughter, Hannah. Hazel held the fishing licences and oversaw the fishing at sea. Henry, who also had a full-time job, controlled the physical aspects of the fishing enterprise. As Hazel explains, they divided up the other household tasks in the following manner: 'When we come in, there might be something that needs to be repaired, so [Henry] has to go do that while I'm doing the housework, like the washing, the ironing and cooking. If he doesn't have anything, he'll help out with the vacuuming or he'll scrub the floor, just to give me a hand. He is a good help, you know. But if there [are] other things that he has to tend to, one of our vehicles needs something done, or could be something that broke down on the boat that he might have to go back down and fix it, or whatever. Well, then he'll go tend to that so that I can tend here and tend to our daughter and whatever else needs doing.' As the interview progressed it became apparent that although Henry did help out by vacuuming and washing the floor, he did so rarely. And when he did help out it was at her request. Apparently he took responsibility for all the maintenance and repair of the fishing equipment, although she held the fishing licences, and she deferred to him on many issues related to the fishing enterprise. In Hazel and Henry's home, the division of labour within the household and the fishing enterprise still resembled that of the more conventional coastal fishing-dependent households.

Dorothy and Don's household was the most integrated of the coastal households in our study. Dorothy and Don, her husband of thirteen years, shared the tasks related to the fishing enterprise. During the fishing season, she acted as shore skipper while he fished. In the winter months they worked together under Don's direction to repair the boat and gear, and to get it ready for the next season. Although the household remained Dorothy's responsibility, Don did do some household chores without being asked. As Dorothy explained:

> Well actually (laughter), I don't know if I should tell you this because you might want him (laughter). When we get up in the morning, he'll help make the bed, which is strange. I guess a lot of men don't. But he'll help

make the bed. We'll get up. We'll have breakfast as a family together. We'll go downstairs. We'll work til dinnertime at nets. I like to come up, say, a half an hour early to get lunch ready. And usually after supper, if I cook a big meal at suppertime, he'll help clear the table with the girls. He'll help set it. Like I said, everything we do is usually together. He helps me, I help him. If the girls need some help for school, he sits a lot and does homework with them. But no, like I said, it's different than a normal family because we're so involved with each other. He'll do that to give me a break. Because when he's gone away he knows that I'm doing it.

Dorothy did not rely on Don to look after their two children on a regular basis, but she would call upon him to help her out occasionally:

I might say in an evening ... 'Gee, I think I'll go out and visit my mother for an hour. Would you mind putting the girls to bed?' and that's fine with him. He realizes that when he gets out in the boat, hours he's not working, he's got that time to himself. Even when the other deckhand ... is sleeping, he's got that time to himself. He realizes, when I'm home, that I've either got the kids or I've got him coming at all hours, or I've got all the bills and responsibilities of taking care of things on my head. So if I want an hour to myself, he doesn't look at it as him having to do everything then, with the girls or whatever. He knows that everybody needs that time.

Both Don and Dorothy accepted crossing gender lines by doing conventionally male or female tasks as part of helping each other out. They assumed that Dorothy would remain in charge of some tasks, mostly those related to the household and their children, and Don would remain in charge of others, mostly those related to the fishing enterprise. Both spouses had a clear set of responsibilities and a firm power base within the household. Although Don did help out, he did not have equal responsibility for the house and Dorothy did not have equal responsibility for the fishing enterprise.

With the onset of the fisheries crisis, some women took on paid employment as a way to help out. Although this strategy generated much needed income to help offset the economic losses in the fishing enterprise, and reduced women's economic dependence on their husbands, employed wives of coastal fishermen appeared to receive little respite from their double day, especially during the fishing season. Women in the workforce had a triple day of labour – domestic labour, work for the fishing enterprise, and wage labour. For example, Susan, who had been married for seven years and had four young children, took on a full-time job to augment the household income from their fishing enterprise. Besides working for the coastal fishery and at her paid employment, Susan managed all her domestic responsibilities:

> I'm working in a hardware store, just a cashier, and you stock shelves and wait on customers – that kind of stuff. I like it because I'm getting out. But I bit off more than I could chew because it was full-time temporary, for five months. Well, I applied for it on a Wednesday and I got a call Sunday. I had it. So that wasn't a whole lot of preparing, and the family hates it because I'm never home. I leave here at nine-thirty in the morning and I don't get home till six in the evening. And up till five weeks ago I was working six days a week at that, and Friday nights till nine. And I just said I had to have one day off, I was starting to get burnt out. I mean I cook, and clean, and bake, and prepare meals. A week ago Sunday I ... took the skin and fat off eighty-two pieces of chicken I bought on Saturday, and cooked it up. This is what I do on my days off so there's meals prepared, because I've never really taken the time to teach the kids how to use the stove. I mean they know how, like they know grilled cheese sandwich, heat up food, and that, but I've always been the one here to do that, so I still feel I'm obligated to make sure the kids are going to get fed while I'm at work – lunches and suppers. So I'm starting to get burnt out.
>
> I go to bed, ten, eleven, twelve at night. My body hurts by the time I go to bed, because I work before I go to work, then I work when I come home. Usually there's dishes to do, laundry to put away. Like Samatha, she'll bring it in, and fold it and put it in the basket some days if she's in a good mood. If she's not in a good mood I've gotta come home at six o'clock, bring the laundry in, do the dishes, and try to get time where you feel you can sit down and unwind. And the boys are in ball this year, so you're going two nights a week, sometimes three. Sometimes, you know, Saturdays and Sundays. It's pretty busy. I told Stewart once I could settle down a little, I'd help him more. He's looking forward to it.

Susan did it all. Her job started as full-time temporary, for five months, but then became permanent. She continued to balance her responsibilities to her household, her job, and the fishing enterprise. Note that she even promised to help her husband more with his work. She was burning out. She needed help and wanted to rely on her daughter but was reluctant to push. How did she keep going?

Although a number of women in coastal fishing-dependent households spoke of their husbands helping when they were home in the winter, most women still relied on children, usually daughters, for the day-to-day routine things to get done. Women continually stressed that they had to ask for help to get things done. For example, one employed woman left a list of chores on the counter in the kitchen before going to work each day. If a task was on the list, it got done; if it was not on the list, it did not get done. But in all cases, helping out was voluntary. Once the woman left the workforce, her husband's help at home ceased.

Getting Him More Involved

When I asked these women how much domestic work their husbands did at home, I knew I had touched a nerve. It became apparent that getting men to do household tasks often resulted in conflict between spouses.[9] Because wives performed domestic work on the basis of marriage and motherhood and not on the basis of a wage, their experience of domestic labour was as unpaid personal service – just as others saw it. It was viewed essentially as a labour of love, not work. But as Oakley (1974) argued so successfully, domestic labour and childcare *are* work, and it is not restricted to women. While these women wanted to be responsible for the organization of their households, they did not want to do all the housework and childcare themselves. Many of the women I interviewed spoke about their ongoing struggles to get their husbands to help out with basic household chores. We will look at six separate couples – Ken and Karen, Graham and Gilda, Quentin and Queenie, Adam and Amy, Rob and Ruth, and Elizabeth and Eric – to gain insight into how these struggles played out.

Karen and Ken and their three children had been trying to cope with the impact of the restructuring of the deep-sea fishery. Ken worked on a double-crewed vessel. Karen was employed full time to supplement the family's income, but continued to do all the household chores. Karen was exhausted. She had been asking – she said nagging – Ken to pick up his clothes and to mow the lawn, but this had not worked. So she finally decided to use a different tack: 'I just got him to pick up his clothes lately. Like he'd take his clothes off and they'd be in a pile in the bedroom, and he'd come home and say, "Is this washed?" or "Is this washed?" And I finally said, "Did you have it out to be washed?" He said, "No." And I said, "Well then it's not washed. I'm not digging through your clothes any more."'

Now that Ken was home more, Karen felt he should be doing the conventional male chores such as lawn mowing. So she started a campaign to get him to cut the grass: 'I refused to do anything with the lawn. I told my neighbours, "I'm not starting that. If I start that, that's it, that'll be my job, and I'm not doing it. It can grow right over my head." Like he gave our lawnmower to his father, so if he wants to mow our lawn he has to go get the lawnmower and come back and do it. This is the end of June. He's done it once. And I'm not doing it. I'm not. I refuse. Because then that'll be it, I'll be expected to do it. So if I don't want something to become my job [I don't do it]. If I barbecue while he's home, that's it, it's my job. He does that.'

Almost every wife had a horror story of when they left their husband in charge of some common household task. Ken appeared to use his incompetence as a way to avoid taking on domestic responsibilities. As Karen said: 'He's a disaster in the kitchen. He can make coffee. That's it. He can't make toast. He's terrible. [When] my niece was baptized – we'd been only married a few years – I had put a roast in [and] everybody was coming for lunch at the house. I said, "Just keep an eye on those so they don't burn." Well I came home, and the smoke was everywhere. He was sitting reading and the smoke was billowing out around him.' Karen never asked Ken to watch a roast again.

Karen's frustration was not based only on Ken's failure to pick up his clothes or mow the lawn. Karen felt that Ken had reneged on his job as primary wage earner as well as on the other 'manly' jobs he had done in the past, such as barbecuing and mowing the lawn. Moreover, now that Karen worked to help support the family, which she saw as partly doing his job, she expected some help with her domestic work. Within their social rules Karen had a legitimate right to demand more work from Ken, but he resisted.

It is important to see that helping out was a form of exchange of services between individuals – husbands and wives. Gilda, who had been married to Graham for over nineteen years and had two children, made this clear when she described how she and Graham helped each other out: 'No, I never feel that he uses me ... I do what I do so that he doesn't have to do it, and then we can do something else together. Like, let's just say he needs boat groceries. Well I'll go get them because I might want to go to a card game in the night-time, and if I don't go get the groceries, well, he'll have to go and then we'll miss the card game, because he's my partner, right? So we might as well do that, and since I'm working, he vacuums, cooks supper, and does dishes and whatnot, too.' Although Gilda and Graham had not formally negotiated this exchange of services, Gilda knew how the system works. She also felt that the exchange was balanced; that is, it worked in favour of them both. But later in the interview it became apparent that this easy-going exchange had been won through Gilda's continual negotiations with her husband.

Queenie and Quentin had been married for twenty-nine years and had two adult children. Their struggle over domestic labour illustrates yet another aspect of helping out: the act was voluntary, the helper got to define the extent of the aid. Quentin, who was retired, had made it clear what he would and would not do in the way of household chores. Queenie was well aware of these parameters: 'If it's something he can do in thirty minutes or under, he might do it (laughter), but if you look at our lawn, it's more than thirty minutes. He does it once in a while, but he does it only if he feels like it.' Quentin would load and unload the dishwasher, but would not scour the pot and pans. If he spilled something he would wipe it up or use the vacuum if he dropped something on the rug. Queenie had been trying to get Quentin to do the laundry – it was something he could do in less than thirty minutes – but he resisted. As Queenie explained: 'I'd come home and do some [laundry] (laughter). I wouldn't do it all. It just kept building up until my days off. Even throwing the laundry in the washing machine, he doesn't do that. I said, "All you have to do is turn a knob, you know (laughter). Why don't you turn it on?" "I'll get it all screwed up," he'd say "because I would turn the knobs the wrong way" (laughter). He's not mechanical at all. It's just an excuse. I think he thinks if he does it once I might expect him to do it all the time. I don't know, but I kind of think that might be his reasoning.'

In these three examples, we see a struggle between partners. In many cases, spouses felt that if they did a task once it would become 'their job' and they would have to do it forever; that is, the helping out would no longer be voluntary but compulsory. When discussing this issue both spouses spoke about 'helping out,' 'giving a hand,' or 'being asked to help,' language that makes it clear that these activities were not considered voluntary acts on the part of husband or wife. It was also clear that they were seen primarily as either the wife's or the husband's responsibilities. The struggle was getting someone to help out voluntarily.

As we have seen, within this subculture men's and women's roles were clearly defined. Men did not do women's work, and men who did a lot of domestic work were considered henpecked or effeminate. Thus, men played down their participation in chores, particularly conventionally female ones, or they would justify their involvement in other uncompromising ways. For example, one man only hung out the laundry in the dark so that no one would see him. Another man claimed his five-year-old daughter cooked all the meals and he only 'helped.' Other men said they did chores to help out during emergencies or other extraordinary events – a wife was sick, away helping an ailing relative, employed part-time to improve the family finances – and stopped doing them once the 'crisis' was over.

Note how the following interchange between the interviewer and Amy and Adam, a laid-off fisherman just rehired by his company, illustrate both his embarrassment and his attempt to cope with an extraordinary event. The interviewer, using a tape recorder, began by asking Amy what her domestic responsibilities were last year, when she was working full time:

> Amy: Cleaning and everything [are my responsibility]. When I was working last winter he did the cleaning and if he knew what I wanted to have for supper that night ...
> Adam (interrupting): It's still taping. Don't say that!
> Amy: Well you did. You helped out.
> Interviewer: You don't want her to say that on tape? Are you embarrassed? (laughter)
> Adam: No, you got to realize I just like to make a little joke out of something like that.
> Amy: But if he knew what I was going to have for supper that night, it would be waiting for me ready when I got home. I'd tell him at such and such a time to turn this on. All I had to do was change out of my work clothes into something else. So if I was working I knew I wouldn't have to worry about anything. And if I said, here's a bill, he'd go pay it. He'd do that. It's like you get dependent on him, too. When I was working I got dependent on him to do little odds and ends for me.

In this interchange, Adam does not want the interviewer to know that he helped out, or the extent to which he did. He tried to deflect his embarrassment and make his concerns into a joke. Amy's reply, 'Well you did. You helped out,' was a socially acceptable way of describing his domestic work. She made it clear that she controlled the domestic sphere. She also stressed that her husband 'helped out' only while she was working to help out the household. As Amy described Adam's domestic labour and how she depended on him when she had a job, Amy spoke to the interviewer, not to Adam. He might as well not have been there. Now that the extraordinary event was over, things were back to normal. Amy no longer worked and Adam had returned to fishing. Adam's help had been appreciated, but it would not be needed or expected.

Ruth and Elizabeth gave up the struggle of trying to get their respective husbands, Rob and Eric, involved in domestic labour. Some women felt it was their fault that their husbands did not participate in domestic work, because they hadn't trained their husbands to do household chores. As Ruth, a wife of twenty years, indicated: 'No, it doesn't really bug me [that he doesn't help]. I have to sit back and laugh 'cause I have trained him [Rob] and the other one [her first husband] exactly the same way. I did exactly the same thing with the kids (laughter).' Other women saw having their husbands doing household chores as a mixed blessing. Some women frequently complained that their husbands liked to cook but never cleaned up afterwards. For these wives the time spent cleaning up negated any advantage of having him prepare meals. Still others did not like the way their husbands did specific chores, and thus would do the chores themselves rather than get into an argument. Elizabeth was married for almost thirty years to Eric, a semi-retired captain. She explained: 'Well I guess I would have to say I feel that the burden is on me to a great extent. Again, I can take some responsibility in saying, if you're going to delegate things and want people to do it, you're going to have to accept the way they do it – which is sometimes difficult. I found myself perhaps saying, "Oh, that isn't the way I do it," and I know that's wrong, to say that, but that sort of thing does slip out, and then, well, he sort of withdraws and says, "Well then I won't do it." So I know that that's my fault. I should have accepted things like that.' Both Ruth and Elizabeth felt it was their fault that their husbands did not participate more in household chores. But by taking on this responsibility they freed their husbands from any qualms they might have had about helping their wives.

This chapter has explored the nature of the struggles concerning domestic labour in fishing-dependent households. But the women's caregiving and social responsibilities extended beyond the immediate household. The next chapter explores these issues.

CHAPTER FIVE

Family, Friends, Acquaintances

I think that no matter what stage of life your children are in, there's always something to do. Mom finally got us grown up and out of the house, then I started having children. So now there's the grandchildren thing to do. She's very active in all of our lives. She's a very big part of our lives, and I, hopefully, will be the same to my kids – although I don't know if it'll be the same, because Mom had daughters. (Paula)

Women's domestic labour goes beyond household responsibilities and the physical care of dependent children. Their responsibilities and obligations also include the emotional and psychological aspects of caregiving and support within the household (Finch 1989). However, women's responsibilities and obligations change dramatically over the life course, extending as well to other members of their families and across generations. As parents they looked after their young children; as adults they take on the responsibility for their ageing parents. Throughout the life course, demands of husbands, parents, grandparents, grandchildren, and in-laws vie for women's labour, with volunteer work, employment outside the home, and work within the fishing enterprise. Thus women as parents, spouses, and adult children must meet a number of responsibilities and obligations: emotional support, practical assistance, financial aid, housing, and personal care, to name but a few. But these responsibilities and obligations cannot always be met with the help of family or other relatives. Others – friends and neighbours – must be drawn into an individual wife's support network. Responsibilities and obligations to kin, friends, and neighbours are constantly renegotiated so that the nature of support networks changes dramatically over the course of a lifetime.

But some of these social obligations and responsibilities cannot be met solely by an individual's informal support system. Government social services take on some of these burdens, such as caring for the elderly or the unemployed through employment insurance (EI) benefits. However, in times of fiscal restraint, federal, provincial, and community levels of government often change their social policies and may cut back their support of social programs. The economic hardship associated with the fisheries crisis put additional strains on social services such as welfare and what became a revamped employment insurance. As household incomes shrunk, little household money remained for services such as daycare, gardening, lawn mowing, home improvements, or repairs to vehicles, appliances, and other household goods. These economic constraints forced people to become even more self-sufficient, or to rely even more on friends, family, and neighbours for help.

All support networks involve an exchange of goods and services between individuals as part of a reciprocal process.[1] One individual gives a 'gift' with the expectation of receiving a 'counter-gift,' and the process continues with each party trying to keep the exchange in balance. By giving the 'gift' one sets up a debt with the receiver of the 'gift.' If one's debt load to another becomes too great, then one becomes dependent on the lender and thus loses one's independence. These exchanges do not usually involve money, but include exchanges of goods, accommodations, labour/time, and services. 'Helping out,' discussed in the previous chapter, is a form of such an exchange.

Exchanges are either direct or indirect. In direct exchanges one receives the same service as one gives: for example, I babysit your children on Wednesday afternoon and you babysit mine Thursday afternoon. In indirect exchanges one receives an equivalent service but not the same service: for example, I babysit your children on Wednesday afternoon and you help me wallpaper the nursery Thursday afternoon. In another form of indirect exchange, someone receives a gift from one person but pays back another person in the same support network. For example, you receive a stroller from one friend, and when your children no longer need it you give it to another friend with a new baby. All of these exchanges attempt to help other people out, yet they also put the giver in a position of power until the 'debt' can be repaid. Over time individuals develop a history of exchange. Each exchange is negotiated either through open discussion or implicitly from the history of the individuals involved.

Social networks – kin, friends, and neighbours – are dynamic and depend on exchange. In most fishing-dependent households, kin networks, particularly parents and children, form the primary basis for support. Friends' networks complement family networks. Community networks provide the structure for large-scale social support for broad-based activities. Support networks can be integrated to reinforce one another, or some networks can contradict others. For example, friendship networks can be used to buffer the social control exerted by family or community

Within this framework we will now examine how fishing-dependent households, and particularly fishermen's wives, attempt to fulfil their obligations and responsibilities, and we will trace how they utilize their support networks to help them in these endeavours. In order to make the discussion clearer, support systems have been separated into three networks: kinship, neighbour and community, and friendship. This division is artificial. For example, in a small fishing community everyone may be distantly related and your husband's second cousin who lives next door might be your best friend. With this caveat in mind, let's begin by looking at kinship networks.

I Can Always Depend on My Family, but at What Cost?

In earlier chapters we have discussed how, within fishing-dependent households,' kin, particularly parents and female relatives, have helped family members meet their daily obligations and responsibilities. Most of these earlier discussions centred upon childcare and domestic chores, and social activities and support networks which reflected the relatively rigid, gender-based division of labour. At family functions, too, physical segregation of kin occurred: men congregated in one room engaged in 'wharf talk' or 'boy talk' while women prepared meals and talked about family and children in the kitchen. This spatial separation replicated emotional and social segregation in kin networks. Women relied more on their female kin – mothers, daughters, sisters, aunts, nieces, and grandmothers – while men relied more on their male kin – fathers, sons, brothers, uncles, nephews, and grandfathers. Friendship and community networks replicated this same gender separation.

But support by kin went beyond the wife's need for help with domestic chores and babysitting. Families systematically exchange services among members, with each member being aware of the nature of the exchanges. For example, Isabel, whose husband of eight years, Ian, a coastal fisherman, was clear about how her household and her parents' household exchanged services. Her mother frequently babysat her children, and in exchange her husband mowed her parent's lawn, shovelled their driveway, and fixed their car. No one in either household formally negotiated this exchange, but each household recognized the exchange as 'helping out.' Throughout our study, couples' primary level of support occurred between parents and their children, but the degree to which each member helped the other, and what tasks were exchanged, had to be negotiated. As with the nature of all 'helping out,' individuals had to negotiate the nature of the aid; it was not expected nor considered an individual's right.

In their Family Obligation Study in the greater Manchester region of England, Finch and Mason (1993) found that of all kin, parents most frequently helped their children, but they also found that there was strong resistance to the idea that anyone had a right to claim assistance from a relative. They argued that family obligations are negotiated over time: 'The essence of this idea is that people develop commitments over time and in ways which are possibly half-recognized but often not consciously planned. One person helps another out in a crisis and the other then wants to return a favour. Opportunities for doing this may occur easily or they may not. Where a pattern of reciprocal assistance builds up over time, each person invests something of themselves in this relationship and becomes committed to it as a relationship through which mutual aid flows' (1993: 167–8).

Having family nearby and relying on them had its advantages and its disadvantages, particularly that of social control. Oriel, who had been married for over eighteen years, recognized both aspects: 'You learn to keep your mouth shut a lot of times (laughter). Like I don't believe in fighting but I don't let people walk over me either, so there's a lot of times I have to stand up for things. More so in-laws, not my parents, but my in-laws, would butt in but now that's under control so. But it's not that bad. It's got its advantages because you've got built in babysitters and you've got people that you can call on if an emergency came.' Many women spoke about having to assert their independence with their parents and in-laws, especially during the early years of their marriage. But all agreed that they relied on their parents and in-laws and expected their parents and in-laws to rely on them when difficulties arose.

And emergencies did come up. For example, Quinsee's parents, both in their eighties, could no longer manage their own home. Their options included moving into a senior citizens home in town, almost a half-hour drive away and where they knew no one, or finding someone in their own community to look after them either in their home or in their caretaker's home. Quinsee and her husband talked it over with her parents, and subsequently her mother and father moved into their basement apartment: 'They have their own little spot and they're happy as could be. They're self-sufficient. Mom still does all her own housework and cooking and looks after dad. Dad is not too well any more, but he still can do for himself and get around still. We didn't want to see them have to go in a home for special care or anything like that yet.'

This situation solved a number of problems for each couple. Quinsee no longer had to worry about her parents living alone and perhaps facing an emergency alone. She could easily keep an eye on her parents, did not have to travel to town to see them, and could readily help them with tasks they found too difficult. The older couple cut down on their expenses; they lived rent-free. Her mother could still be independent and do her own housework and other chores, but her responsibilities were reduced. They continued to live in their own community and to interact with their friends and relatives. Moreover, Quinsee's mother also helped out: 'So if the ironing stacks up she loves to iron, she'll do that (laughter). If she doesn't have anything to do some days she'll come up and, "Got anything for me to do?" So that's good. It's nice to have help, that's for sure!' Both couples felt that they had reached a mutually acceptable agreement.

From negotiations such as these, responsibilities and obligations developed over time. In a moment of crisis one person helped out and then the other person wanted to reciprocate. For example, Rosemary's mother had been ill, and, as the only daughter, the work of caring for her mother had fallen on her. In fact she had given up her job to help out. Although Rosemary was close to her two brothers, it was her sister-in-law who gave her the most support: 'My sister-in-law, she's been a very strong support for me in the last couple months, very much so. She's been a godsend. She knows I'm there for her, anytime.' Although Rosemary's sister-in-law helped her to look after her mother, her sister-in-law's greatest aid was the emotional support she provided by listening to Rosemary's worries and fears concerning her parents' health. Rosemary recognized her debt to her sister-in-law and expected to be called upon in the future. At the same time, Rosemary's brothers were indebted to her for her services, and she expected them to help her in future times of crisis.

The nature of responsibilities and obligations within all families is dynamic. Where mothers and mothers-in-law used to help the young wives and mothers, now these women help out their parents and parents-in-law. Later in life siblings and their spouses look after their ageing parents, and, in some cases, their grandparents. For example, some of the women in the study talked about specific negotiations they had with their siblings about the care of their parents. In one case, the siblings drew up a schedule listing who would go to the hospital to visit their ailing mother, who would look after their father, and who would look after the parental home. But these same 'children' were also looking after their own children, and babysitting and caring for their grandchildren. For the middle-aged members of the family sandwiched between their parents and their children, the obligations and responsibilities became enormous. Some, like Rosemary, decided that they could no longer do it all. Some quit their jobs. Others brought in hired help. Others renegotiated their contribution vis-à-vis other family members.

Throughout the life course other aspects of family networks contracted. Parents got old and died. Brothers and sisters moved away or died. Children, nieces, and nephews moved away and/or started their own families. As Elizabeth, a woman in her fifties with a grown family, explained: 'My family is pretty well diminished. I have a brother who lives in a nearby community, and my husband has a brother and a sister, but we don't have a large family network anymore, and I sort of miss that. I guess if I'd lived somewhere else, perhaps I wouldn't have missed that, but I do miss my mom a lot, and his parents as well.' Many women attempted to compensate for this decline in family contacts by relying more on friendship networks and community.

However, kinship networks were not simply loci for exchanges of services. Each member of a family had developed a reputation. Were you seen as needy, uncommitted, or unreliable? Were you seen as dependent or independent? Finch and Mason (1993:170) saw the dependence-independence theme as 'a strong theme in our data ... that most people strive to ensure that they do not become "too dependent" upon assistance of relatives. A certain degree of interdependence is allowed, indeed is highly valued. But situations where one person receives more than they can return to the other is definitely to be avoided.' In my study individuals also attempted to keep their commitments and obligations in balance. For example, at the start of their marriage, Pat and Paul, who were married for over twenty-three years, wanted to be independent of their parents. They felt that their parents had enough to do raising their younger siblings, so they tried not to call upon them for much help. As Pat said: 'We got some support from our parents, but our parents both have big families so we didn't really like to rely on them a lot because we felt they needed a break. My youngest brother was just getting out and off, so it didn't really feel like the minute he left home that I could say to my mom and dad, look after my children today or tomorrow, because I felt they needed a break from it. So we tended to have babysitters, or stay home quite a bit, not go out too much.'

One result of being dependent on family support was the possibility of social control. Relying on others can both increase and decrease stress (Pearlin and Schooler 1978). For example, if a wife relied too much on her mother-in-law for babysitting, she could create an obligation for the husband towards his parents, which had both emotional and social costs. It could leave the wife vulnerable to social control by both her husband and his parents. One of the most common forms of social control in these households had the mother or the mother-in-law refusing to babysit on occasions that they deemed inappropriate for a married woman. In one case a woman said that her mother-in-law, after refusing to babysit for her, threatened to tell her son what his wife was doing – going to town alone to see a movie – if she found someone else to babysit for her. In another case, a wife's mother reluctantly babysat her first grandchild, although she thought her daughter should stay at home, and refused to babysit at all once the second grandchild was born. The daughter had been placed in a situation where she had to find another babysitter she would have to pay, or stay home with her children. Both social pressure and financial considerations led the daughter to give up her job. She stayed home until both children were in school, and then returned to work.

But social control exerted by the family can enhance some individual situations. Laurie came from a farming family, while her husband Len's family always fished. Laurie found the first few years of marriage difficult. She and Len had financial difficulties arising from their coastal fishing enterprise. When she spoke to her own family about these problems they were unsympathetic – if you marry a fisherman what do you expect. However, her mother-in-law was extremely supportive. She ex-plained to Laurie how the fishing enterprise worked and what was expected of her. She also taught her specific skills she needed. Her in-laws also talked to their son about Laurie's concerns. Ultimately Len curtailed his coastal fishing and began to fish mainly deep-sea.

Living close to parents or parents-in-law also had its advantages and its disadvantages. Claire and Colin, married only two years with one small child, rented an apartment in town while they fixed up a trailer located

Table 5.1 Place of Residence at Marriage for Coastal and Deep-Sea Samples (in percentages)

Place of residence at marriage	Coastal sample	Deep-sea sample
With her parent	10.8	14.7
With his parents	18.2	12.7
On their own	70.9	72.7

Missing values (0)

on a piece of land adjacent to his parents' and sister's homes. During these renovations Colin's parents had been very helpful by looking after their small daughter and aiding in some of the work. However, Claire had some misgivings:

> I'm surrounded by in-laws (laughter). I mean I get along good with them and everything. But I know once we're moved into the trailer I'm going to have to set a little bit of ground rules, or something like, call before you just pop in type of thing, because I know they're going to (laughter). We could be running around naked or something (laughter), and they just 'knock, knock.' I like being close to them, but I don't want to be like right up their arse (laughter). I don't know. I never thought about it until after I had the trailer up there, and we're up there working and it's like constantly when we're down in the back room doing some work, they'll just come right in type thing. I don't want that once we're moved in there. I mean, I don't do it at their house, so I wouldn't want it done at mine.

Setting ground rules, and negotiating responsibilities and obligations, are key to any relationship, but were essential for Colin and Claire. The potential disadvantages were overcome by the advantages of having their own home and land.

Living next door to one's parents created particular strains on relationships, but living with one's parents after marriage put even more strain on newly-wedded couples. In our study, just over a quarter of all newly-weds lived with either the husband's or wife's family (see Table 5.1). In the coastal sample almost 11 per cent of couples lived with her parents, and a little over 18 per cent lived with his parents. In the deep-sea sample almost 15 per cent of couples lived with her parents and a little less than 13 percent lived with his parents. This phenomenon

Table 5.2 Distance from Kin for Coastal and Deep-Sea Samples (in kilometres)

Kin relationship and distance from fishing household	Coastal sample	Deep-sea sample
Wife's parents * **		
0 to 10 km.	57.1	58.5
more than 10 km. to 50 km.	21.8	23.2
more than 50 km.	21.1	18.3
Wife's siblings * ***		
0 to 10 km.	40.6	39.0
more than 10 km. to 50 km.	22.3	20.9
more than 50 km.	37.1	40.1
Husband's parents * **		
0 to 10 km.	79.1	65.3
more than 10 km. to 50 km.	11.8	14.0
more than 50 km.	9.1	20.7
Husband's siblings * ***		
0 to 10 km.	56.9	40.1
more than 10 km. to 50 km.	17.9	22.5
more than 50 km.	25.2	37.4

Missing Values *(0); **(1); ***(5)

reflected economic necessity rather than preference. Many deep-sea fishermen's wives felt that living with their parents or parents-in-law gave them some respite from loneliness, but at a social price. By the end of the first year, most couples moved out of the parental home.

Proximity to family members explained much of the differences in kin-based relationships among households. I assumed that the nearer one lived to kin the greater the probability of interacting with them. In the survey I asked women how far they lived from their parents and siblings, their parents-in-law, and their husband's siblings. These results, summarized in Table 5.2, broke down, by distance, into three categories – ten kilometres or less, ten to fifty kilometres, and more than fifty kilometres – representing driving time of ten minutes or less, ten minutes to an hour, and more than an hour.

Most of these fishing-dependent families lived physically closer to the husband's (patrilineal) kin than the wife's (matrilineal) kin. Nearly 80 per cent of coastal families and 65 per cent of deep-sea families had the husband's parents living within ten kilometres of his home. About 57 per cent of coastal families and 40 per cent of deep-sea families had husbands' siblings living within a ten-minute drive from his home. Wives' parents and siblings in both the coastal and deep-sea fisheries lived further away. About 58 per cent of women's parents, and 40 per cent of women's siblings lived within a ten-minute drive from their home. In the Newfoundland literature, patrilocality of fishing-dependent households characterized the coastal fishery (e.g., Faris 1966; Stiles 1971). Patrilocality remained strongest in Nova Scotian coastal families.

Newfoundland studies have indicated that men most frequently fished with their patrilineal kin. Nova Scotian studies to date have not shown this characteristic.[2] In this study, one deep-sea fisherman fished with his brother on a company boat. Of the forty-four helpers[3] in the coastal fishery sample, two reported fishing with their sons, one with his brother, and the other with his wife. Of the 107 captain/owners, fifteen men fished exclusively with someone from their household. The remaining ninety-two captains employed in total 117 persons, and approximately half (60) of these helpers were kin. Of these relatives, two-thirds were patrilineal kin (16 sons, 15 brothers, 4 fathers, and 6 cousins), one-sixth were wives (10), and the remaining were affines (5 brothers-in-law, 1 son-in-law, and 3 fathers-in-law). Thus approximately one-third of helpers were patrilineal kin of the captain-owner. This finding could reflect a household and family strategy to spread income within the family during tough economic times. It may represent an older pattern of men working with their siblings or fathers.

Not having family nearby created a different set of problems. For example, Anne, married to Abraham with three children, had no family to rely upon. Anne's family lived in Central Canada and she infrequently saw them. She found that her friends who had family nearby did not appreciate their advantages and her difficulties. As she said:

> Sometimes when I hear people whining or complaining about family matters – I'm really intolerant of it. I think you should be more grateful that you have your family around you, and you shouldn't snivel because you get to take your kids there a couple times a week. I really lose my patience with people that complain sometimes. And yet on the other hand, when you do have a lot of family together, maybe there's more fighting. Maybe it's not as ideal. Sometimes I wish I had more family, especially at occasions – Thanksgiving, Christmas, birthday parties. Those are the times that I really wish they were around. But for the most part, I think I do okay.'

But as we saw earlier in the chapter, relying on family had a cost either in the form of social control or of as yet unreciprocated debts. Generally households with little or no family nearby relied more on their other support networks – friends and neighbours. So what did Anne do?

> When I move into a town – I've moved about four or five times – instantly I get a babysitter. I don't have any family. It seems like everyone has aunties and uncles, and I don't have that. So I find babysitters. So I guess I do rely on babysitters. And when I do meet new moms, sometimes you can ask them to reciprocate babysitting, things like that.

Many women, like Anne, relied on friendship networks to substitute for local kin networks.

My Friends Help Out

Friendship networks of fishing-dependent families resembled family networks in many ways, even though individuals could choose their friends. Negotiated exchange of goods and services between individuals also defined these friendship networks. Debts and counter-debts built up between friends and a history of exchanges developed. Let's return to Anne's discussion of how she and her friends shared childcare responsibilities: 'Well, you take care of mine [children] this weekend, this day or something, and then we'll just switch it around on another day. I do have a couple of friends that I can call on the night before and say, look I have an opportunity to go into Halifax, can you take the kids at lunchtime? And I don't like to abuse it. Like, I think it's something that should be done just once in a while. But I'm fortunate, I can call on several friends around and get help that way.' These relationships between friends were constantly being renegotiated, and dependency would shift back and forth between individuals. Note how Anne mentioned not wanting to abuse her friends. Most friendship networks in fishing households were gender-based or couples-based. But unlike kinship networks, friendships sometimes ceased when the relationships no longer offered the support they once did.

Friendship networks are dynamic. Part of the caregiving role that women take on includes social activities which relate directly or indirectly to their responsibilities and obligations inside the home. Each type of friendship sets up different forms of exchange which relate to the specific responsibilities and obligations these women have within their households at any particular time. In the study, at the beginning of marriage – 'when they were new' – a woman's friendship and support network revolved primarily around family and friends associated with the fishery, the school, or the workplace. When a woman had a child, her support network expanded to include other new mothers in her neighbourhood, while many of her single friends, and friends from work and school, fell away. Once her children were in school, she had new demands on her time, such as the local parent-teacher association (PTA) or lunchroom duties, which offered additional opportunities for her to meet women in a similar situation. Children's lessons, sport activities after school, or church activities such as bake sales, gave her even more chances to meet other mothers with youngsters about the same age and with similar interests as her own. If she was employed, fellow employees might also be included in her friendship network. In later years, as her children grew up and left home, she might participate in volunteer work at the local hospital or community centre, in service clubs, or in church groups, thus expanding her friendship network.

Fishermen's friendship networks appeared to be drawn from their friends from school, their home community, and their wives' support networks. For the few men who had worked outside the fishery, workmates from previous jobs might still be part of their friendship network. But for most fishermen, especially deep-sea fishermen, other fishermen they worked with or had worked with in the past dominate their friendship networks.[4] When their husbands came ashore, wives attempted to integrate their husbands into their expanded and diversified social networks. Incorporation into a 'couple network,' which reflected their wives' dynamic friendship network, allowed these men to break away from their occupationally dominated support system. Many wives saw this integration as part of the 'taming' process, meant to make their husbands as happy with the 'shore world' as they were with the world of the sea. As we saw in an earlier chapter, not all men made this transition.

But some women, especially those who were older and/or were employed, often had another circle of friends who had no connection to the fishery. For example, Holly and Harry, both in their mid-forties, had a wide range of friends. As she explained: 'We have a circle of friends that are fishing friends, and then we have another group who are people that have other types of jobs. Not because they're fishing people, and it's just that it's a different circle. You tend to know a different group of people from the fishing line, and then other people from your other social life. And I guess through me being out, socially active, too, and then within the church you find people also ... but yes it is a mixed group. But never the two groups together. I don't know (laughter), it's just never ever happened.' Segregation of fishing from non-fishing friends seemed to be common in these households, reflecting the segregation within the communities themselves. But through their community involvement or employment, women bridged the gap between fishing and non-fishing friendship networks.

The commitment of fishing-dependent households to the wider community and their integration into it translated into a variety of activities for family members, such as community service, school and church programs, and social clubs. Through these activities women from many different economic and occupational backgrounds came together. Out of these interactions friendships emerged, and other social activities not directly related to women's household obligations developed. As the fishermen's wives' friendship networks developed and expanded, wives became integrated into the social networks of their new friends. But many of these activities outside the household might compromise a woman's fulfilling her familial roles as wife, mother, and daughter by setting up competing demands.

Sometimes, however, social networks not only failed to give an individual the support she needed, but might work against her. Family, friends, and community exerted social control over their members. For example, we discussed earlier how some newly married fishermen's wives felt that their mothers/mothers-in-law spied on them and reported their activities back to their sons/sons-in-law. Once these women had children, they felt that their mothers and mothers-in-law restricted babysitting to what they thought were socially acceptable events (e.g., bingo, school activities). Community members also controlled behaviour through gossip and by refusing to help out except for socially acceptable activities.

Women broke away from such constricting codes of behaviour through their friendship networks. Donna, a young mother, explained that by developing friendships with other deep-sea fishermen's wives she escaped some of this control: 'You [fishermen's wives] pretty much just hang together – come over to my house for supper, or go to their house for support. You got five women there that look like they're single, and then you've got the guys coming over (laughter). "Sorry but we're married." "Well, where are your husbands?" "Well, they're fishermen." And then you get some of them that think, well, he's at sea so what do you care, right? And I guess some go for that.'

Both deep-sea and coastal fishermen's wives employed this strategy.

Although it was a generally positive strategy to cope with loneliness, support networks such as Donna's might also create their own special difficulties, particularly if the wives frequented restaurants and bars when their husbands were away for long periods of time. Some women who did engage in these activities complained that local men would phone them up and badger them to go out with them, knowing that their husbands were away. The more conservative members of the community labelled them 'lounge lizards.' Husbands got reports of their wives' activities – no matter how innocent – and jealousy and fears of infidelity abounded. Community and family members viewed the 'misbehaviour' of fishermen's wives as an indication of their husbands' inability to control them. This perception itself led to conflict between spouses. For the wives, socializing together formed a joint act of rebellion, where each woman supported the other in dealing, after a fashion, with the common problem of being a fisherman's wife. Some wives felt that these actions helped them cope with their particular problems. Yet they paid for this freedom with social disapproval.

Another area of social control involved community attitudes to physical abuse within marriage. These attitudes varied from zero tolerance to feigned ignorance. Women in abusive situations called upon family, neighbours, and friends to aid them. However, leaving a relationship was difficult at the best of times, but extremely difficult in small communities with conservative family values and few, if any, social services. Many women had slight financial resources, and what resources they did have were tied up in their household finances. Of five women in our study who left abusive first marriages, all indicated that it was their friendship network, not their family, who gave them the support to leave. This support ranged from emotional support encouraging a wife to leave, to helping her to pack and move, say, to Halifax, to helping her get a job, to lending her money to set up a new apartment.

Friendship networks also helped people enhance their work experience. For many women, getting along with co-workers, talking and socializing with people at the workplace, and seeing these same people outside business hours opened up new opportunities. Many fishermen's wives took the opportunity of developing friendships with co-workers and through these friendships expanded their couples network. These friendships, unlike those developed with other fishermen's wives, cut across occupational, class, and community boundaries.

Just as friendships at work opened up new opportunities, so did volunteer work, community service, and sports. Many women explained

Table 5.3 Indicators of Wives' Belonging to Community for Coastal and Deep-Sea Samples (in percentages)

Indicator	Coastal sample	Deep-sea sample
Wife feel that she ...		
really belongs	57	50
belongs	32	41
not really belongs	11	9
knows their neighbours	72	48

Missing values (0)

that volunteer work led to paid employment through friends whom they met at their unpaid work. Others spoke of meeting women who shared similar interests, which allowed them to try new activities. For example, encouraged by her friends, one woman, who enjoyed singing in the church choir, tried out for a part in a musical put on by the local theatre group. She got the small part and has gone on to larger roles, becoming part of the theatre crowd.

I Can Call on My Neighbours in a Pinch

Research on Nova Scotian fishing-dependent households has indicated that coastal households have had a strong sense of community, while those households associated with the deep-sea fishery did not appear to be as closely tied to their communities and more closely resembled the rest of Nova Scotia (Apostle, Kasdan, and Hansen 1985; Davis 1985). I used two measures as rough indicators of how integrated their households were in their communities: (a) if they felt that they belonged to that community; and (b) did they know their neighbours. I followed up this inquiry in the individual interviews. Wives of coastal fishermen felt more embedded in their communities. They had lived longer in their communities than had deep-sea fishermen's wives – on average twenty-one years, versus seventeen years, respectively (see Table 5.3). Data on their husbands showed the same pattern. Coastal fishermen's wives were more likely to know their neighbours and to feel really part of the community than were their deep-sea counterparts.

The port from which a deep-sea fisherman sailed affected the choice of the community where his family lived. This choice of residence profoundly influenced his household's way of life, especially when he was at sea. In larger centres, opportunities for leisure activities were more diverse, and the more cosmopolitan the community in which one lived, of course, the wider the range of socially acceptable activities. Larger communities offered social services ranging from childcare facilities to transition houses, from health-care clinics to employment offices, as well as job opportunities. But some fishermen's wives found these communities impersonal and had difficulty making friends. They complained they knew no one and that their children went to school with strangers.

The choice of community also affected the deep-sea fisherman's social contacts. The nature of his work made it difficult to maintain or develop social contacts beyond his shipmates. If his family maintained its residence in his home community he had to travel, in some cases many hours, to reach home. But when he got home, he had his extended family and friends for support. If his family took up residence in the deep-sea port, he had the convenience of having his family right there, but he had no social network of his own except for his fellow crew members. This lack of social and community interaction increased the fisherman's sense of alienation. It also put additional pressure on wives not only to develop their own social network but also to develop a couples network.

In the case of fishermen originally from Cape Breton or Newfoundland, who now fished out of Lunenburg, on the coast of Nova Scotia, some of their families had followed them to Lunenburg, while others had remained in their home communities. If they stayed in their home communities, many problems emerged. The wives of these fishermen seldom saw their husbands. A number of fishermen residing in Lunenburg had families living in Newfoundland, whom they saw only two or three times a year – when they took a specific trip to visit home, or when the family came to visit them, during the vessel's refit, or over Christmas. These long-distance relationships put additional stress on their families, although the wives usually had a strong support network in the home communities to meet the problems of everyday living.

More frequently, especially for those fishermen who had seniority and some job security, their wives and families moved to the deep-sea port. This solution allowed the wife and family more time with their husbands/fathers, but it meant they were isolated from their extended family and friends, and had to adjust to communities in which they were strangers. Most of these women developed support networks with other women in the same situation, or with wives of men who worked on the same vessel as their husbands. Commonly, they also relied on community resources – church associations, clergy, social service workers, doctors, and other medical services – to a greater extent than did other fishermen's wives.

Many deep-sea fishing-dependent households, especially when they came from a coastal fishing community nearby, decided to reside in these smaller communities, rather than the larger centres, so the wives could have the support of family and friends while the husbands were away at sea. Coastal fishing-dependent households tended to live in small, and in many cases isolated, communities in closer contact with family and other kin than deep-sea fishing-dependent households. Most coastal fishermen fished in the bays and coves within view of their homes.

Again, living in a small community had both advantages and disadvantages. Everybody knew each other. Children easily visited their grandparents or played with their cousins. Most of the adults had grown up in the area, they knew the parents of their friends, they knew their neighbours, and they were willing to help each other out. As Amy said, 'Usually if anything bad happens, a house fire or anything like that, everybody just jumps right in and helps.' People knew when someone was ill. They knew when a car broke down. They knew when someone needed a lift to town. And most were willing to help. In these communities, gossip worked as a conduit for information, telling community members the area's news.

But here as well there was often a price to pay. These communities had few social activities except bingo, playing cards at the fire hall, joining church-sponsored events, visiting friends' homes, and participating in family gatherings. There were few social services. In times of stress, such as when a relationship turned sour, women found it difficult to get professional help, and they had to travel for support to transition or halfway houses located in Lunenburg, Halifax, Liverpool, Shelbourne, or Yarmouth. Furthermore, community solidarity and social expectations further limited women's options. In many of these small communities social pressure often forced women to stay at home despite abusive relationships. Even in more normal times women's activities could be curtailed through social pressure. For example, although Amy generally enjoyed living in a small community she saw that it had a few drawbacks: 'Everybody knows you. They might know when Adam would go away, so if someday you decide to go visit somebody and you're not home until about one o'clock, well they've got you doing God knows what with God knows who. And you're not doing it. So there, all of a sudden, you've got major big problems. And if no one knew you, they wouldn't give a shit.'

Almost all the women I interviewed who lived in these small communities complained about their stifling nature and their rigid code of behaviour. Many of these women felt that some community members saw their attendance at certain social events as inappropriate for the wife of an absent fisherman. Forms of acceptable entertainment included social outings with their immediate and extended family or their husband's, and bingo,[5] church socials, and home and school functions, usually in the company of other women or family members. Going to town to see a movie, to eat at a restaurant, or to do other activities by themselves or with other wives was unacceptable. Wives felt that if they deviated from these norms, they were the focus of gossip by their neighbours. Petty jealousy and feuds abounded. Some people had not spoken to some of their neighbours for years.

Thus if you wanted to be part of the community, you had to make compromises and attempt to get along. You had to conform. In these communities, where everyone knew everyone else's business, gossip acted as a major instrument of social control. As Virginia stated, 'If you do something wrong, you can be sure everybody's going to know that you've done something wrong, so you just gotta learn to live with that part.' Not everyone had this laid-back approach. Others, like Gilda, found the gossipmongers

> awful – oh the backbiting and backstabbing of some of them. Some people give you snide remarks. I don't do that and I don't like it done to me, it hurts my feelings. [It is mostly] about your children. Like nobody has a perfect child. [They] all do bad stunts. I'm sure I did them, you did them, or whatever. But they like to tee-hee and haw-haw and give you digs about your child. But if theirs does anything, you never hear about it. I think it's jealousy. They're jealous or something. I don't know what it is. But I try not to get caught up in that. I just stay away from it. I've got my own little boys here that'll do what boys do. You have to worry about what goes on in your own house, not somebody else's, right?

Gossip maintained the rigid code of behaviour and conformity in these small communities. Almost everyone I interviewed had a story about how gossip had touched someone in her household.

Family, Friends, Acquaintances 109

Living through gossip, especially false gossip, was difficult. For example, when Oriel went on vacation and Obadiah, her husband, started fishing out of a port further down the shore, a rumour started that they had split up. Although false, the rumour travelled like wildfire through the community. Then Oriel started getting phone calls from people at work asking her about her break-up. The rumour went through the company grapevine as well. So Oriel spent a whole week of her vacation trying to straighten that out. How did she get through it?

> (laughter) I've lived through this. It's not the first time. It won't be the last. I don't know how to say it, it's just you learn to take it like water off the duck's back and you just let it ride. You just say, well if they talk about me they leave someone else alone. But in that case I had to answer phone calls. Like the phone was ringing off the hook from my friends saying, God, so and so said that you'd broken up. And I said, no I haven't. I don't know why they would even think this. Well it's all around in your neighbourhood, you better straighten it out. I'm not going to. Not in this neighbourhood, forget it (laughter). So I called my husband, told him. He goes, oh well, look at it this way, they're talking about us, they're leaving someone else alone. And that's our attitude. That's the only way you can deal with it. But for my first week of vacation it was a bit bizarre, you know, getting phone calls like that.

Oriel and Obadiah's healthy attitude sustained them through this ordeal and they could laugh about it. Their case was not unique, and not everyone came out of it as well as they did.

In the interviews I asked women about their neighbours and the kind of interaction they had with them. For example, did they rely on their neighbours for help? were they close friends? Few women appeared to have neighbours as close friends, relying on family in the neighbourhood instead. Most appeared to agree with Ina and kept their neighbours at a distance, calling upon them for help only occasionally. As Ina said:

> I have good neighbours, like I'm not one of these that likes somebody to be dropping in every so often for a cup of coffee. But if I run into difficulty, which I have many times when he [Ivan] was out to sea, I have a good neighbour down here on this side of me, and all I have to do is go down. If it's some problem I can't fix on my own, or don't know what to do, I just go down and I ask him. But we're not, you know what I mean, we don't run

back and forth all the time, but I know where they are if I need them. I have a lady friend down the road, and me and her, we do go out for lunch when [her husband] is out aboard the boat. Or sometimes – she has a daughter in the city – I will go to the city with her.

Most interactions with neighbours appeared to be more businesslike in nature. In time of crisis women would rely on neighbours for help. Common interactions with neighbours included borrowing equipment or supplies; hiring the neighbour's son to cut the lawn, shovel the snow, or do other household chores; and asking neighbours to look after children for short periods of time. Neighbours with young children about the same age tended to have more interactions and had more friendship-like relationships. These women most often shared cooperative childcare and spent time visiting their neighbours. But in general people appeared to be wary of owing 'social' debts to their neighbours, and of becoming too dependent on them. Again, the group of women that did rely on neighbours the most were those who did not have kin in the area. Women with family or in-laws there had ready access to kin support networks.

But in the 1990s populations in rural coastal communities had declined and a number of new problems had emerged. With the fisheries crisis many homes had gone up for sale, and had been sold as summer residences to people from Germany, the United States, and other parts of Canada, or as commuter homes to people working in Halifax and vicinity. Other families had left the communities without selling their homes, leaving the houses vacant with a relative looking after them. The exodus of young people and middle-aged couples with their families left many communities overpopulated with older couples. Without their families for support, these couples had to turn to the few remaining residents for help when they needed it. Some communities no longer had the critical mass of year-round residents needed to support community activities, such as the volunteer fire brigade, children's and adult's sports teams, or Girl Guide and Boy Scout troops. Lack of activities and opportunities for children, especially in the teen years, meant a lot of free time for children or long drives to the larger centres where attractive activities were offered.

The influx of strangers, especially during the summer months, ameliorated some of the problems associated with lack of critical mass in these communities, but it produced other problems, such as how to integrate these newcomers into the community. Virginia saw the problem this way:

They're just transients. It seems a lot of times they just move in and out. You see their cars flying by. They work in town and they have their own circle of friends that come to the house. I mean, you say maybe hi on the road. We used to make an attempt. There were certain people that used to move in, when we were younger. You'd go with a casserole the first time they moved in, and try to make friends. And some of it worked. But other people, like I said it's so transient. They may be here six months, eight months, a year, and then they're gone again. So you can't make friends that way. It's just that a lot of the older people's passed away and they put younger people in their homes. You don't get to be friends with them.

But the transient nature of newcomers was only one problem. For example, how did one get these newcomers to participate in and commit to community projects?

The influx of newcomers brought a clash of values. New community members wanted different amenities from the people who lived in the community year-round and had for most of their lives. For example, the newcomers did not see the need to refurbish the community wharf to be as important as changing the zoning in order to build a restaurant. Newcomers saw the community wharf with its stored tubs of smelly baited trawl as a public nuisance, and they wanted the tubs banned. For many tourists, summer residents, and commuters these communities represented picturesque fishing villages, precious remnants of the past – commodities to be consumed. To long-term residents, these communities still remained viable commercial and residential fishing areas struggling for survival. These two uses of space conflicted, and they set the two groups against each other in competition for ever scarcer resources.

In this chapter we saw how women's social networks sustained them by allowing them to meet their obligations and responsibilities to themselves and to their families. In the next chapter we will see how these same networks gave a social meaning to the lives of these women.

CHAPTER SIX

Just Having Fun

I was just thinking about this the other day, we do not have very many friends outside of family, and it's a big family so that kind of takes care of most of our social needs. In the community itself we have friends that we are friendly with. We may go to their house for a visit once in a while, but mostly it's within the fishing community. (Trudy)

In fishing-dependent households women's leisure involved squeezing time from their double if not triple day of labour (Berk 1979; Chambers 1986; Harrington, Dawson, and Bella 1992; Shaw, Bowen, and McCabe 1991; Wellman 1985, 1999). This meant grabbing a few quiet moments after the children went to school to have a cup of coffee while scanning the local newspaper, or watching the soaps while doing the mending or ironing, or chatting with a friend in her home while knitting lobster heads for traps, or going to town with a sister to do the weekly grocery shopping, or rushing out with a workmate for a quick lunch. Like so much of their lives, a woman's leisure was curtailed by the demands of her domestic labour and paid employment, by the work schedule of her spouse and other members of the household, by her children's needs, and by her commitments to other family members, friends, and community.[1]

By far, most of women's leisure activities took place with family members either in their home or within the community. Most women spoke of playing with their children, watching them engage in sporting events, and taking them to the park, library, cinema, or swimming pool. Many women saw having a quiet evening at home watching TV or a video, playing cards or board games, or having family and friends drop by, as a break from their regular routine. A few women played sports, went regu-

Table 6.1 Frequency of Wives' Involvement with Various Forms of Entertainment for Coastal Sample When Husband Is at Home and Away (in percentages)

Form of Entertainment	Husband away			Husband home		
	Never	Infrequent*	Often**	Never	Infrequent*	Often**
Video store[1]	52.3	38.3	9.4	44.7	45.3	10.0
Watch sports[2]	51.3	20.1	28.6	50.8	21.7	27.5
Play sports[3]	66.4	18.8	14.8	64.2	20.3	15.5
Take children to library[4]	73.2	24.1	2.7	73.2	24.1	2.7
Go to library for self[1]	70.5	28.2	1.3	70.0	28.7	1.3
Take children to movies[4]	73.2	26.8	0	58.9	41.1	0
Go to movies[1]	79.7	20.3	0	60.0	40.0	0
Play bingo[1]	62.4	26.1	11.5	62.0	27.3	10.7
Go to a restaurant[1]	32.9	57.1	10.0	2.0	81.3	16.7

Missing values [1](1); [2](32); [3](3); [4](39)
**Infrequent = less than once a week
**Often = once a week or more

larly to the gym, or exercised in their homes. Socializing outside the home frequently meant going out to a movie or a restaurant with their husbands or friends, or going over to another couple's home for a meal or drinks. (See Tables 6.1 and 6.2 for a summary of the frequency of some of these activities.)

The differences in individuals' responses related to personal preference, the husband's work schedule, the availability of disposable income, and the closeness of facilities (Donaldson 1991; Glyptis 1989; Glyptis and Chambers 1982; Morris 1990; Rubin 1994; Wheelock 1990). When their husbands were home, wives of coastal fishermen watched more videos, took their children to the movies more often, went to the movies more themselves, and ate at restaurants more frequently than when their husbands were fishing. In some cases husbands participated in these activities; in other cases they looked after the children or took on other household responsibilities so their wives could go out with her friends or their children. In general deep-sea fishermen's wives participated more in their children's activities when their husbands were away than their coastal counterparts did, but their involvement decreased

Table 6.2 Frequency of Wives' Involvement with Various Forms of Entertainment for Deep-Sea Sample When Husband Is at Home and Away (in percentages)

Form of entertainment	Husband away			Husband home		
	Never	Infrequent*	Often**	Never	Infrequent*	Often**
Video store[1]	47.0	33.5	19.5	46.3	28.2	25.5
Watch sports[2]	32.5	23.6	43.9	38.6	21.9	39.5
Play sports[3]	68.3	11.0	20.7	69.2	10.2	20.6
Take children to library[4]	74.8	21.5	3.7	74.8	21.5	3.7
Go to library for self[5]	79.5	19.1	1.4	82.2	16.4	1.4
Take children to movies[4]	52.3	45.8	1.9	54.7	44.4	.9
Go to movies[5]	74.0	26.0	0	65.1	34.9	0
Play bingo[1]	73.2	13.4	13.4	70.9	15.6	13.5
Go to a restaurant[6]	28.0	52.0	20.0	9.3	61.4	29.3

Missing values [1](1); [2](36); [3](5); [4](43); [5](4); [6](0)
*Infrequent = less than once a week
**Often = once a week or more

when he came ashore. As we know, when husbands were home from the deep-sea fishery, their needs became paramount, and activities with them were privileged.

Entertainment venues in small communities are often limited to community or fire halls, churches, schools, and sports facilities. In many fishing-dependent communities these facilities are the loci of leisure activities, where card games, bingo, dances, socials, sporting events, concerts, and plays take place. Larger centres, such as Lunenburg and Bridgewater, have additional social venues such as restaurants, bars, shops, cinemas, and libraries. With more disposable income available, deep-sea fishing-dependent households went to restaurants more frequently, and were more likely than their coastal counterparts to go to sporting and concert events, the theatre, and dinner in Halifax.

Whether in small communities or one of the larger centres, community organizations such as the PTA, service clubs, church organizations and youth organizations are the social backbone of rural Nova Scotia. In these communities men's interests tend to revolve around political issues such as the building of new roads and wharves, volunteer associa-

Table 6.3 Frequency of Wives' Involvement with Community Organizations for Coastal Sample When Husband is at Home and Away (in percentages)

Community organisation	Husband away			Husband home		
	Never	Infrequent*	Often**	Never	Infrequent*	Often**
PTA[1]	34.9	63.3	1.8	33.3	64.9	1.8
Church services[2]	36.9	41.6	21.5	36.0	42.0	22.0
Choir and vestry[2]	84.6	10.7	4.7	84.7	10.0	5.3
Sunday school[2]	90.6	1.4	8.1	90.0	2.0	8.0
Service clubs[3]	85.2	12.1	2.7	85.9	11.4	2.7
Girl Guides/ Scouts[4]	84.7	3.0	12.3	86.9	3.0	10.1
Leader for Girl Guides/Scouts[5]	91.7	5.1	3.1	91.7	5.1	3.1

Missing values [1](43); [2](1); [3](2); [4](52); [5](55)
*Infrequent = less than once a week
**Often = once a week or more

tions like the firemen's brigade, and sports such as curling, golf, or hockey. It is common for women to pour their labour into programs for their children, such as Girl Guides and Boy Scouts, the PTA and sports leagues; and into community events including church, school, fire hall socials, and fund raisers, as well as service clubs and volunteer work. Women in this study participated in a variety of social activities available in their communities, including the PTA, church services, choir and vestry activities, teaching Sunday school, service clubs such as the Kinettes and Lionettes, and Girl Guide/Boy Scout troops. These activities, although not a comprehensive list, give some indication of the breadth of their participation (see Tables 6.3 and 6.4).

For wives, participation in community activities did not depend critically on their husbands' fishing schedules. What did make a difference was each family's life cycle. For example, Zoey used to teach Sunday school, and belonged to the church guild before she had her three children. Once she had a family, she dropped out of her church-related commitments and became involved in activities at the local school. Once the children were older she joined the firemen's auxiliary. This type of revolving participation was a common pattern. When women had children in school, they participated in child-focused activities. (Of course, women who were employed at this time in their life cycle found it difficult to put much time into these activities.) Once their children

Table 6.4 Frequency of Wives' Involvement with Community Organizations for Deep-Sea Sample When Husband Is at Home and Away (in percentages)

Community organization	Husband away			Husband home		
	Never	Infrequent*	Often**	Never	Infrequent*	Often**
PTA[1]	28.0	63.6	8.4	28.0	62.7	9.3
Church services[2]	40.3	37.6	22.1	45.0	34.2	20.8
Choir and vestry[3]	89.7	8.2	2.1	89.0	8.9	2.1
Sunday school[4]	88.4	5.5	6.1	88.4	5.5	6.1
Service clubs[4]	94.6	4.8	0.7	95.2	4.1	0.7
Girl Guides/ Scouts[5]	65.7	2.0	32.4	67.7	5.0	27.3
Leader for Girl Guides/Scouts[6]	83.7	10.1	6.2	83.7	9.2	7.1

Missing values [1](43); [2](1); [3](4); [4](3); [5](51); [6](52)
*Infrequent = less than once a week
**Often = once a week or more

had left home (or before they had children), women participated in volunteer work, service clubs, and church activities. In general, deep-sea fishermen's wives participated more often in activities with their children and less frequently in community-based activities than did their coastal counterparts.

All of these activities contributed to the well-being of the community. Without the voluntary labour needed to run them, many of these activities would not have been possible. For example, if mothers did not do 'luncheon duty' – supervise children in the lunchroom at the local school – the children would have been forced to go home during the noon-hour break. Many of these children had mothers employed in the workforce who had no alternative arrangements for their children's lunchtime care, and supplying childcare over the lunchtime would have posed an extra drain on their finances. Other forms of volunteer labour included working in the local retirement homes, reading to the elderly, or making in-home visits to them. The volunteer fire brigade ensured a service the government did not provide to rural communities.

Small communities thrive on their social events. These events are not only opportunities for socializing, they also raise money for community improvements which would otherwise not be funded. Community events also create a venue for social exchange. In rural Nova Scotian communities, 'socials' such as dances, barbecues, church suppers, and other fund-raising affairs are widely advertised, and participation cuts across all age, gender, religious, and class boundaries. Being responsible for activities such as an ice cream social at the local school sets up one form of social debt; partaking in these activities establishes another debt. Community organizers have two expectations: first, that all members of the community will come out and support an event; second, that members of the community who partake in these activities will at some point take on some responsibility for the event. These responsibilities fall disproportionately on women, who commonly make foodstuffs, cook meals, serve food, and sell tickets for events in their communities.

Going Out and Staying In

For most fishing-dependent families socializing took place in the home,[2] or involved visiting with their family or friends, or going out to the movies and restaurants. But opportunities for these activities were constrained by the husband's livelihood. As Trudy, a coastal fisherman's wife explained:

> There's no time [for socializing]. You get up at three o'clock in the morning, like in the spring when he's fishing outside by himself, he gets up here at three o'clock and I may not see him till any time that night. I may see him at two o'clock [in the afternoon] if there's not a lot of fish. If there's a lot of fish I may not see him till seven or eight o'clock [in the evening]. When you live a life like that, you don't feel like socializing (laughter), you're just glad to stay home. So we don't do a lot of visiting outside of 'necessary visiting,' which is pleasure, too, but not to just deliberately go and say well tonight we're going to go somewhere. I think being away from home so much, home becomes your place for relaxation and pleasure.

Many coastal women I interviewed shared Trudy's attitude about socializing during the fishing season. During the off season they had more opportunities to go out and see folks. Much of couples' socializing was called 'necessary visiting,' a local term which included going to see family, visiting a sick friend in hospital, participating in rite of passage events such as weddings, funerals, wakes, and graduation, and seeing other fishermen about business while wives chatted over tea in the kitchen.

Many deep-sea fishing-dependent households followed a different pattern. Socializing as couples formed a basic component of the husbands' being home. It included 'necessary visiting,' usually as a couple, getting out and seeing friends, shopping, and just doing things together. Developed over twenty-nine years of marriage, Queenie and Quentin's visiting patterns illustrated this point:

> I'd say most of our visiting is [with] relatives. He has his set of friends in the daytime, I guess, and I have my set of friends that I go to see. There aren't too many couples that we visit, but there are a few. It's just occasional and not regular visits, but he's not one for going out and visiting much. If he does, it's not for very long. He'll go for an hour at the most, and then, 'Now I'm ready to go home' (laughter). Usually we just stay at home and have sandwiches and coffee, or whatever – finger foods. Sometimes a barbecue, this time of year. They [his friends] might drop in for something, for a reason, but never to sit and visit, or stay or anything. They [her friends] come to see me, usually when he's not at home. But if he is home, I mean, they don't wait till he's gone to come, but they often know (laughter) when he's out so that's when they usually come for the evening. Usually just sit and drink coffee and talk (laughter). Sometimes we go out for supper, or go shopping together, my friends and I.

For most households – coastal or deep-sea – there were three distinct groups of friends who visited: his friends, her friends, and their friends (couples). Most individual friendship networks segregated along gender lines, but there was also a couple's network where the membership of the two networks might overlap.

In deep-sea fishing-dependent households, particularly among captains, mates, and other officers, socializing became more elaborate with formal dinner parties among other captains and officers, or among friends outside the fishery. These couples also went out to see and be seen. In some cases this meant going out to fancy restaurants along the shore or in town. In other cases, couples enjoyed going to concerts, theatre, or dinner theatre in Halifax. Elizabeth and Eric's social activities illustrate this point:

> We go out to friends [homes], we do that together. We go out to dinner a fair amount. I'd like to share a few more things with him, but I think, as a fisherman, it's difficult to get him involved. I find, as I get older, that I'm doing a few more things on my own. I've started to go into Halifax to some theatre and things like that, and it's a little difficult to get him involved. He

Table 6.5 Comparison of Activities with Other Fishing Couples for Study Samples (in percentages)

Activities	Coastal sample	Deep-sea sample
Dinner in home	66.1	71.4
Vacations	26.1	25.4
Cards	68.7	76.2
Drinks	53.9	68.0
Going out	65.2	78.6
Number of cases	115	126

enjoys dinner theatre, so maybe some day I'll get him to *Cats* or something like that when it comes. Again, we can share things like golfing, because it's purely social and recreational that you do it for. It's easy to do that with him periodically and to go to friends and barbecue, that sort of thing.

In some households, like Elizabeth's, the wife organized these activities as a way to pull her husband out of his 'wharf' world. In other households, husbands saw these activities as proof of their success and so encouraged their wives to participate.

Many fishermen and their wives socialized with other fisher couples, and these friends played an important part in their social network. In the study I asked women if they counted other fishermen's wives among their friends. Eighty-three per cent of deep-sea fishermen's wives and 76 per cent of coastal fishermen's wives said yes. These couples frequently had dinner in each others homes, shared vacations, played cards, engaged in social drinking, and went out in the evening together (Table 6.5 summarizes these data). In all of these activities, except shared vacations, deep-sea fishing-dependent couples had more contact with other fishing couples.

Making friends with other couples who participated in the fishing industry had many advantages. For example, most of Trudy and Tim's friends worked in the coastal fishery and did business with Tim. The catalyst for their socialization in the first instance involved his work. 'It's Tim needing to talk to one fisherman, so we'll go, or they'll coming here in the evening, and then they'll do their talking.' This form of men's 'necessary visiting' created the opportunity for friendships to develop between their wives. Once the women became friends, the couples' socializing was less work-oriented, and the women started doing things together without their husbands.

In the deep-sea fishery, socializing among fishing families correlated with the various crews the husband sailed with throughout his career. Wives found that by incorporating his 'family' at sea into home social events, they could have more control over his evening activities (possibly modifying his drinking patterns) – especially important during the early years of marriage. By making friends with wives of other fishermen sailing with their husbands, they also had a group of women they could rely on when their husbands were away.

As noted earlier, both coastal and deep-sea fishermen spent much of their leisure time engaged in 'wharf talk' or 'boy talk.' This included chatting at a local restaurant, 'having a mug up' (drinking coffee) at their homes, or while repairing gear at the wharves, or before sailing. As Brenda described these sessions, 'It's not a business discussion, it's not a business conference. It's just a social sort of thing. They may not talk about fishing the whole time, they may talk about somebody (laughter), but the majority of the time it's fishing, cars, things that they're interested in.' Gender segregation remained strong within the fishing communities. As Faye described her and her husband Frank's visiting patterns: 'Socializing with fishermen, other fishermen, that's fine. He can go down to the wharf and talk all day to other fishermen, or other men can come here and they can sit and they can talk for hours about fishing. But that's basically it. Like that's all it is. When the other wives do come here, we usually go in the front room and we have our chat. The men are out here having their chat (laughter). So it's not too much socializing, like, men and women together. It's like Frank has his spot; I have my space (laughter).'

Women more frequently had 'non-fishing' friends than did their husbands. These networks usually developed out of women's relationships with other women whom they worked with either in volunteer activities or paid employment, whom they met through their children or some of their other activities. But many women did not develop these friendships beyond socializing at work or supporting each other with their children's activities such as babysiting, or carpooling. (See Table 6.6 for comparison between fishing and non-fishing friends.)

A few women, like Susan, had developed friendships outside the fishery. Susan had formed a social network of four or five women, whom she described as her 'gang': 'We sit around and gab, drink pots and pots of tea, and smoke our heads off. There's basically four or five of us. We

Table 6.6 Comparison of Women's Activities with Other Fishermen's Wives and Workmates for the Study Samples (in percentages)

Activity	Other fishermen's wives		Workmates	
	Coastal sample	Deep-sea sample	Coastal sample	Deep-sea sample
Shopping	53.0	61.1	50.0	42.6
Talk on phone	91.3	94.4	68.7	73.1
Babysit	45.2	36.6	14.5	9.0
Car pool	22.2	15.9	6.0	1.6
Go to restaurants	45.2	58.1	79.7	71.6
Drink with	19.1	29.8	35.9	34.3
Go to bingo	47.8	37.1	23.4	11.9
Social clubs	28.9	16.8	17.2	12.1
Number of cases	114	124	64	67

chum around. And all our kids are in school now, so we can run, like, up to the grocery store, or the hardware store, or sometimes we'll go out for lunch.' All of Susan's friends had paid work except one, and none of them was married to a fisherman. Susan complained that they had very little time to see each other because of their family and employment commitments. Although they tried to include their spouses, these attempts failed because the 'guys have nothing to talk about.' The development of a 'couples network' that does not include people involved in the fishing industry breaks down the gender and occupational segregation discussed earlier; however, only a few women appeared to have successfully integrated their husbands into these networks.

Having a Few Drinks

Many social activities in fishing communities included drinking alcoholic beverages.[3] In the survey I asked women how often they participated in a range of activities and whether the activities involved the consumption of alcohol. I also asked those women who drank alcohol on these occasions how often they did so (Table 6.7 summarizes these results). It should be noted I was asking women about their drinking habits and not those of their husbands. Women most frequently drank at social occasions: attending a party, social gathering, or wedding; having friends or relatives to their home, or going to friends' or relatives'

Table 6.7 Frequency of Women's Drinking While Participating in Various Social Activities in the Year Preceding the Survey (in percentages)

Activity	Coastal sample*		Deep-sea sample**	
	Never drink	May drink	Never drink	May drink
Spend a quiet evening at home	59.6	41.4	54.8	45.2
Spend time at someone else's home	39.7	60.3	41.5	58.5
Have friends or relatives visit	33.8	66.2	39.3	60.7
Go to a restaurant for an evening meal	50.3	49.7	45.9	54.1
Go to a restaurant for lunch	89.4	10.6	81.5	18.5
Go to a bar or tavern	56.3	43.7	50.4	49.6
Go to a club or meeting	86.1	13.9	94.8	5.2
Engage in outdoor leisure activities	59.6	40.4	53.3	46.7
Engage in sport activities	92.7	7.3	96.3	3.7
Attend a party, social gathering, or wedding	21.2	78.8	30.4	69.6
Go to a concert, sport event, or festival	95.4	4.6	91.7	8.3

Missing values *(2); **(15)

homes. Coastal fishermen's wives appeared to consume alcohol more often on these occasions than did deep-sea fishermen's wives. On the other hand, deep-sea fishermen's wives imbibed more frequently when they went to restaurants or bars than their coastal counterparts did. Deep-sea fishermen's wives also went out to these venues more often. This pattern may reflect the differences in disposable household income and/or access to licensed locations. Over 40 per cent of women reported consuming alcohol while engaged in outdoor leisure activities, such as a barbecue at the beach, a picnic, and camping, or while spending a quiet evening at home. In these cases, deep-sea fishermen's wives reported drinking more alcohol and more often than did their coastal counterparts.

Social drinking took place in fishermen's homes, lounges, taverns, restaurants, and other social forums. Some unmarried fishermen, especially those involved in the deep-sea fishery, spent their whole shore leave 'drinking with the boys.' The participation of married men in these heavy drinking sessions depended on their integration into their family and community, and their wives' attitude to drinking. Most wives felt that social drinking among fishermen should be curtailed. Janet, whose husband Joe and her father both drank heavily, made the following comments about the situation:

The husband comes home, the first day you don't see him. They're at the tavern drinking. I think it's just a way of life. I think it's just that they've gotten into the routine of it, and they don't want to stop it. I mean, I've known guys who come in, the boat lands at noontime and they're in the pub till five or six instead of going home. If you have a family and you have been away for – well trips used to be two weeks – if you've been away that long, and it's no big deal to you ... What's more important ... your family or your beer (laughter)?

And it seems to me a lot of the wives say, well we don't mind. We don't mind? If they don't mind why are they talking about how much they don't mind? If they don't mind, why isn't it just forgotten about? You don't talk about it if you don't mind something. It's no big deal to you. But it's, 'We don't mind. I don't mind.' No, I mind (laughter). I always minded. I can remember having to go down for Dad. Mom and I'd sit in the car, and we lived [down the shore] and the boat was in Lunenburg. We'd sit in the car ... until I finally had enough and said, 'I'm going to go get him. I'm not going to sit out here and wait any more' (laughter).

Many of the women I spoke to about their husband's drinking began their statement with 'I don't mind him drinking,' but by the end of their statement it was clear that they, like Janet, minded very much. Many women had similar experiences to Janet's, but involving a different male relative: sometimes it was their father, and other times it was their husband, their brother, or their son.

Fishermen drank both at home and away from home. Table 6.8 compares wives' reports of their husbands' drinking patterns inside and outside the home. (The data includes only those fishermen who reportedly drank.) With respect to coastal fishermen, wives reported that approximately 30 per cent of their husbands never drank at home, 64 per cent never drank outside the home, and only 6 per cent drank in both venues. With respect to deep-sea fishermen, wives reported that approximately a third of their husbands never drank at home, 42 per cent never drank outside the home, and a quarter drank at both venues.

Whether fishermen consumed alcohol inside or outside of the home, both venues presented problems. Most wives wanted their husbands to spend time with their families when they came ashore. Social drinking, especially when wives were excluded, encroached into family time. Arguments surrounding the amount of time spent drinking could lead to serious conflicts. As Janet explained:

I have the feeling when he's home four days he shouldn't be drinking, that

Table 6.8 Comparison of Coastal and Deep-Sea Fishermen's Drinking Patterns Inside and Outside the Home (in percentages)

Frequency	Husband drinks at home		Husband drinks outside of home	
	Coastal sample	Deep-sea sample	Coastal sample	Deep-sea sample
Never	29.7	33.8	64.1	41.8
Once a year or less	11.7	8.1	10.9	11.9
Several times a year	21.9	25.0	10.9	19.4
About once a month	5.5	12.5	6.3	11.9
Several times a month	14.8	7.4	4.7	6.0
Once a week	9.4	3.7	2.3	3.7
Several times a week	7.0	9.6	0.8	5.2
Number of cases	128	137	128	137

he should be spending time with his family. Instead of that, he goes out with his friends. Of course, the first day home, everybody's around and everybody's got a bottle. It's the same old story all the time. He knows pretty well sometimes when I'm angry, too. I find that the worst part of it is that if he has a few drinks then he loses all sense of time, and you're spending all your time waiting, waiting, waiting. Waiting for him to come home, waiting for him to decide to sit down and eat, waiting for this and waiting for that. And if you make plans, he comes about two hours later, well you know how you feel – teed off. I guess it's pretty well the same for a lot of people.

Many men tried to get their wives to drink with them. They felt that by 'co-opting' their wives they could avoid arguments. But some women, like Janet, did not drink themselves, and this led to increased tension and arguments. As Janet explained: 'I tend to avoid anybody that does drink and he [her husband Joe] don't get mean when he gets drinking, he gets silly (laughter). And I get angry (laughter).'

Drinking at home had one major advantage: 'If they're drinking at home at least you know where they are.' Here is how Donna, a young wife and mother, described her husband's drinking patterns:

If he's stressed out in the house around me, he's just cranky: 'Don't talk to me.' I don't look at him, or anything like that. He just likes to drink when he comes home, but he's not, like, drunk every day or anything like that.

When he's gone so much, he really doesn't have a whole lot of friends. He does have one friend that lives [nearby], and if he's stressed out, instead of having a drink with me he'll call him and say, 'Come on down and have a drink.' It annoys me sometimes. I mean, I'm here. Why can't he have a drink with me (laughter). It's a male bonding thing (laughter).

Consuming alcohol at home sometimes set up petty jealousies. For example, Donna not only felt rejected because Dean phoned a friend to come over to drink with him, but she also felt neglected when Dean's friends and family came to visit when he was home but ignored her when he was at sea. As Donna explained:

When he's gone, I get no visitors. It's, like, people know Dean's home and the yard is just full of cars, and it drives me nuts. Like there's always somebody here, whether they're in the house or outside with him, it seems like there's always somebody here. That gets on my nerves, but then, like he says, he can't do a whole lot about it because what can he say, 'Get lost?' But I say, 'Who's more important, your friends or me?' (laughter) and he just doesn't answer. Like, come see me when I'm here alone, right? But no, I guess they don't like to come and visit me because they're his friends (laughter). Even family, they don't seem to come around as much, and then when Dean's home, they're always dropping in.

Gradually, as the years go by, most husbands' drinking patterns changed. Most men matured, decreased their drinking, and drank less frequently in public places. As Queenie, married to Quentin for over twenty-eight years, said:

He used to drink too much. He stopped drinking hard liquor about ten years ago, and maybe for five or six years he didn't have any alcoholic beverages. But now he's gone back to beer – only beer – and he's keeping that under control pretty good I would say. Before he stopped he did drink too much. When he'd come home drunk he'd just be sleeping all the time, unconcerned about any of the rest of us in the house. I never bugged him (laughter) because I think I tried that once and it only made things worse. So I just said, 'No, I'm going to stay out of this.' What made him actually decide to stop, I don't know, but maybe he got so sick (laughter) once on one of his hangovers that he just decided it was time. I don't know. He just stopped. I guess he just came to his senses (laughter) and stopped. So, since then things have been going very well. I'd say, no problems. It wasn't

Table 6.9 Frequency of Fishermen and Their Wives Engaging in Binge Drinking (in percentages)

Type of drinking	Coastal men	Deep-sea men	Coastal women	Deep-sea women
High-risk binge drinking*	7.1	21.1	1.6	0
Lower-risk binge drinking**	37.8	24.2	22.7	21.9
Binge drinking	44.9	45.3	24.3	21.9
Number of cases	127	128	128	123

*High-risk drinking is four or more drinks three or more times a week.
**Lower-risk binge drinking is four or more drinks at a drinking session but less frequent than three times a week.

good with him drinking, with the children growing up. There were times when he embarrassed us, and he embarrassed the kids when they had their friends in and things. But there was never any violence or anything like that. Well, anyway, that's all in the past (laughter) now. At least ten years he's been, except for the beer, everything under control, so I'm glad (laughter).

But some men never changed their drinking habits.

Binge drinking, or going on a tear, was the most common problem for fishermen's wives. Table 6.9 indicates the frequency of binge drinking among coastal and deep-sea fishermen and their wives, as reported by fishermen's wives over the last year prior to the survey. Binge drinking means consuming four or more drinks at a single drinking session. High-risk drinking implies binge drinking on three or more occasions over one week. (A drink was defined as one bottle of beer or glass of draft, or one glass of wine or wine cooler, or one straight or mixed drink with one ounce and a half of hard liquor.)[4] Approximately 22 per cent of all women who drank reported low-risk binge drinking behaviour. Two per cent of coastal fishermen's wives reported high-risk binge behaviour. Wives reported that approximately 45 per cent of their husbands had engaged in binge drinking behaviour. Approximately 21 per cent of deep-sea fishermen and seven per cent of coastal fishermen engaged in high-risk binge drinking.

Sixty-nine per cent of the wives reported that their husbands had ever gone on 'a tear' or a single bout of binge drinking. Of these men 73 per cent had gone on at least one tear during the year prior to the survey while

6 per cent were reported to have gone on a tear more than once a month. For about 6 per cent of the deep-sea fishermen, frequency of binge drinking meant going on a tear virtually every time they came ashore.

These drinking binges took many different forms. Some men became silly when they drank, some said things they could or would not admit to remembering in the morning, some men stayed home, drank themselves into a stupor, and slept it off. Others went off drinking with their buddies, and their wives never saw them during that shore leave. Some men turned violent, got into brawls, or turned on their wives and children. Others smashed things, drove their cars recklessly, went hunting, or engaged in other risky activities while drinking. This type of behaviour became a source of accelerating tension between spouses, and, as his drinking problem started to affect others (e.g., brawls), or property (e.g., crashing a car), or involved the police (e.g., drunk driving), wives' level of stress increased. But no matter what form these binges took, the women did not directly intervene. Many fishermen's wives comments were similar to Isabel's: 'I try to ignore him now when he's drinking. There's no way to deal with somebody that doesn't know what they're saying. There's no way to deal with somebody like that when they're drinking, so the next day I try to sit down and talk to him. When he's drinking I just try to ignore him. Let him try to sleep it off.'

The consequences of others' drinking impinged on the lives of most fishermen's wives. In the study both the survey and the in-depth interviews identified a number of these problems (see Table 6.10). Problems affecting fishermen wives included having serious arguments, being humiliated, having marital difficulties or family difficulties, being disturbed by loud parties, being pushed, hit, or assaulted, having friendships break up, having property vandalized, and having financial troubles. About 18 per cent of these women reported being a passenger with a drunk driver and 2 per cent reported being involved in accident involving a drunk driver. Table 6.10 compares the frequency with which the wives of deep-sea and coastal fishermen experienced problems caused by other people's drinking in the preceding year, with the experiences of Nova Scotian women as reported in the National Alcohol and Other Drugs Survey.[5] In all cases, save 'being disturbed by loud parties,' the deep-sea fishermen's wives (column two) reported experiencing these problems with greater frequency than other Nova Scotian women did (column three). Similarly, in all cases, save 'being disturbed by loud parties' and 'having friendships breakup,' the deep-sea fishermen's

128   Set Adrift

Table 6.10 Comparison of Consequences of Other People's Drinking on Women Based on Data from Both This Study and the National Alcohol and Other Drugs Survey (in percentages)

Activities	Coastal sample	Deep-sea sample	NADS – Nova Scotia women*
Serious arguments	30.9	36.7	16
Being humiliated	29.5	36.0	22
Marital or family difficulties	18.1	25.3	8
Being disturbed by loud parties	34.9	22.0	26
Being pushed, hit or assaulted	4.7	8.7	7
Having friendships break up	12.1	9.3	7
Having property vandalized	4.7	4.7	3
Having financial troubles	2.0	2.7	2
Passenger with a drunk driver	17.4	18.7	10
Accident because of a drunk driver	1.3	2.0	1
Number of cases	149	145	702

*National Alcohol and Other Drugs Survey, Eliany et al. (1992: 272–3).

wives (column two) reported experiencing these problems with the same or greater frequency than did coastal fishermen's wives (column one).

What could one do about a husband's drinking? The experience of Elizabeth, whose husband Eric frequently went on binges, illustrates how many fishermen's wives approached the situation: 'I guess I have my own coping mechanisms – I enjoy the time now when he's away because I just don't deal with that at all. And when he's here (sigh), there's a bit of dread if I know he's out and he's drinking because each man is different in how he reacts with drinking. I have a friend that just goes to sleep and I always say, 'Why can't you be like that?' (laughter) He's not. I always say he's the ever-ready battery, and he just goes on, and on, and on, until he finally, sort of, drifts off. He's somewhat argumentative, you know. I really try to avoid that type of thing.'

Although Eric could turn violent he seldom did. When he did turn violent, Elizabeth turned to her friends for support:

> I've never gone to Al-Anon or anything. I just never seemed ready for that, although I've had a friend, a dear friend, who encouraged me to do that, but I just never have. It might have helped me! I don't know. Well I guess I do. I suppose it's had its affect on me, but I seem to have gone through it

and coped with it. I mean, I have friends who have similar problems, so I suppose we share that. You know you're never alone with something like this. Ah, it would be nice if it just got a little bit more in perspective in life, and just didn't dominate at times. I suppose now, and even in the past, I find it interfered with social aspects. If we were trying to have a good time together, the alcohol could always cause it to turn out to be a bad event instead of a good event, and I regret those things.

Like Elizabeth, many women whose husbands engaged in binge drinking talked about how a good time can go bad. Many of these women talked about how they felt responsible for their husband's drinking, felt they should have been able to stop them and get them home. Elizabeth continued: 'It just can interfere with social activities. And now, today, I often wonder, well, when we go, why is it that we had to come home with one person so drunk, or why did I have to be responsible for getting him home, you know, that sort of thing. I mean, I just don't want to be his keeper in that way. He's old enough and big enough to take care of himself, but yet I protect him from driving, I protect him from whatever. I take care of him. And it's a little hard to take after many years. You just don't want to do it anymore. I guess I took on that role, so I'll probably have to continue that (laughter).' Elizabeth also felt that at some level she must shoulder part of the blame for her husband's activities. It was as though her drinking produced his behaviour: 'I mean I do drink myself, and I'm not without my flaws there. I mean, I know that alcohol makes me more vulnerable to being quick, and taking things personally, and so forth. Of course maybe I misjudge him there, do you think maybe I do?'

When drunk, some husbands' behaviour led to violence. For example, Yasmine's first husband became physically and mentally abusive when he drank. Because there were no facilities in the small community where they lived, Yasmine would have to go to one of the larger centres almost an hour away for treatment and counselling. At first, Yasmine went to her doctor. 'I was a mess a couple of times that I went there, like mentally [as well as physically], and they put me in touch with a counsellor. And I went to see her a couple of times and she just let me talk, and I knew what I had to do, but it was getting the courage to do it.' But it took Yasmine a long time to decide to leave her husband. She found that she needed more support in order to get up the courage to leave. She turned to AA and Al-Anon for help: 'Basically the only thing around here is AA and Al-Anon ... I even tried that. But there isn't enough support groups; the nearest is an hour away. And I mean a lot of women

Table 6.11 Reported Frequency of the Type of Help Given Fishermen's Wives Related to Husbands' Drinking (in percentages)

Type of help	Coastal sample	Deep-sea sample
Call family member	14.6	17.2
Call friend	10.4	14.1
Call neighbour	2.1	4.7
Call police	4.2	12.5
Call minister/priest	0	0
Take him to a doctor or hospital	16.7	14.1
Been taken to a doctor or hospital	2.1	6.3
Leave home and go to family, friends, or neighbour	20.8	21.9
Go on retreat	4.2	4.9
Pick him up at police station	2.1	10.9
Number of husbands who go on tears	48	64

probably want to go, but to go that distance they've got to tell their husbands something. They wouldn't want them really to know where they're going.' Finally, Yasmine did leave her husband after her children left home. She regretted that she waited so long, but felt she had no community or family support. As she said, '[Now] they're talking and really saying what it's like and how bad it is. Before, if a woman got beat up by their husband they wouldn't say anything, and now they do. They don't lie and say, well I ran into something.'

As indicated in Table 6.11, when their husbands went on tears women had few options. Most stayed home and waited for their husbands to return, others avoided or ignored them. But sometimes these strategies did not work and the wife needed help, so they called on a family member, a friend, a neighbour, or the police. Some women took their children and went to friends and relatives, or even on religious retreats. Some reported picking their husbands up at the police station; some reported taking their husbands to a hospital or doctor for alcohol-related reasons. Sometimes the men turned their violence towards their wives and families. Six per cent of deep-sea fishermen's wives and 2 per cent of coastal fishermen's wives reported going to a doctor or a hospital for treatment for alcohol-related injuries.

When it got so bad they could no longer tolerate their husbands' drinking, some women issued an ultimatum: 'Stop drinking or I'll leave.' As Karen explained:

I grew up with my father drinking. Mom never drank unless Dad was home, but when my dad was home, she was drinking. Ken's an ignorant, mean, mean drunk. But my mom and dad weren't like that, but still they were drinking, and not themselves. And so, I just said, 'I just can't go [on like this.]' When you grow up with it, you don't want to live with it forever, indefinitely. So, he stopped. I think he must have wanted to, too. I don't think he could've stopped if he didn't want to. I don't think anybody can make you. I don't think anybody makes you drink, and I don't think anybody makes you stop. If you want to do it, you do it yourself. But I still wouldn't live with it. If he started drinking again, I'd be gone. Three kids or thirty kids, I'm not doing it. Friendly drunk or non-friendly drunk to me is a drunk – it doesn't matter.

But some of the women were not willing simply to wait for their husbands to grow up and take responsibility. These women took a more aggressive approach to their husbands' drinking and either left themselves or told their husband to leave.

Other women just gave up fighting. They felt defeated by their husbands' behaviour and believed that they could do nothing to change it. But they also believed that they could not make a clean break. Opal's husband, Oscar, had a drinking problem for most of their twenty-four years of marriage. At the time of the interview Oscar had just been charged with drunk driving. 'I'm conditioned to it,' Opal explained. 'I've learned to be independent for myself. And sometimes it's like he's just a nuisance. Maybe it was too much stress in my life. I like the peace and quiet – no hassles, no nothing. And it's always a hassle, always something every year, something to aggravate. I don't know, but I am at the point [that I] don't want no more hassles ... with this Breathalyzer charge, I was very upset, and it's like, what else is new? Every year it's always a disappointment, because we should've been able to retire.' Opal's frustration over Oscar's drinking soured their relationship, yet she saw no alternative than to stay with him.

Drinking, Stress, and the Job

For most fishermen consumption of alcoholic beverages on shore played an important role in socialization and the relief of stress.[6] Both coastal and deep-sea fishermen's wives emphasized that their husbands drank in part because of stress generated from fishing; however, the underlying reasons for the stress were different. Coastal wives talked about the pressures of making a profit, of finding enough fish, of worrying about how the government could change the quotas or other fishing policies. These financial concerns were similar to those of other small-business owners. Deep-sea wives talked about the alienating working conditions on board, the companies' management practices and exploitation of the limited stocks, and the government's policies towards the fisheries. Their focus was similar to other blue-collar workers in industries in financial difficulties. Both groups cited uncontrollably dangerous working conditions.

Sometimes the high levels of stress associated with this dangerous working environment overwhelmed the fishermen's psychological mechanisms for coping. Three common manifestations of mental stress at sea were 'the channels,' 'going on a tear,' and 'breaking up.' 'Breaking up' took many forms. In extreme cases fishermen suffered mental breakdowns. For example, during my previous study of health and safety concerns of deep-sea fishermen, one man calmly packed his duffle bag, told the captain he was going home to speak to his wife and would be right back, and then walked off the stern carrying his bags. More commonly, 'broken' men simply became 'nervous': they appeared edgy around machinery, would worry that the boat would sink, wore their survival suits all the time, or seemed anxious while shooting away or hauling in the nets.

'The channels' usually occurred after an extremely stressful trip, such as one with a major fire or an accident to a shipmate, but they also occurred after a series of trips involving only moderate levels of stress. Most fishermen attributed 'channels' to being too long at sea and needing a break. Symptoms varied. When sailing for home, afflicted fishermen felt restless, impatient to get ashore, and could not sleep. When they came ashore, they found it difficult or impossible to adjust to shore life; they stayed awake and might drive for miles going nowhere in particular. An afflicted fisherman might go to a bar or the Legion Hall, and drink until he passed out, or talk about trying to catch up or being all 'broke up.'

However, the most common manifestation of stress involved 'going on a tear' or on a drinking binge. For some men, usually single or separated, this binge drinking could become the crux of their life, indicating their alienation from their work at sea and perhaps resentment that this was all there was to life.

For most, 'being on a tear' remained an occasional occurrence. Nevertheless, wives whose husbands were experiencing this much stress found greeting their husbands at the dock a trying experience. The organization and working conditions of the Nova Scotia fisheries, particularly the deep-sea fishery, combined with the stress inherent in working at sea, provided a structural framework for these drinking patterns and an environment that supported high-risk drinking behaviour. Both fishermen and their wives spoke of these men when they came ashore as being 'like animals let out of a cage.' As Gail, the wife of Gordon (a deep-sea fisherman) said, 'I honestly believe that the fishing industry has created more physical abuse, more alcoholism, [and] drug abuse' than any other industry. But whether officer or crew, the intense work experience of the deep-sea fishery in particular, with its gruelling conditions that cut men off from family and friends for long periods of time with minimal breaks between trips, made the employment situation of fishermen extreme among blue-collar workers. The additional stress associated with the fisheries' decline of course exacerbated the situation, and women were supposed to mediate these emotional and psychological effects. Indeed, their wives' labour enabled most of these men to work in these conditions. Wives worked hard to maintain their families through caregiving and social activities, thus creating a positive alternative to the drinking culture. Their success was marked by the fact that single or separated men were the ones least able to cope with the stress of their jobs.

Just as these women sustained their households and family members through their domestic labour, caregiving, and social networks, many also contributed to the financial well being of their households through paid work. The next chapter will examine the characteristics of these women's employment and the role it played in sustaining their households.

CHAPTER SEVEN

Going to Work

After working for seven years full time, you sort of think, oh gosh, a two- or three-week vacation – you sort of count the days. But, I mean, that's part of life, and work is really what makes life satisfying. My family has been important to me, and I wouldn't trade those years at home, but to a certain degree I think that going to work was really good for me. (Elizabeth)

Women's paid employment outside the home has become the norm among Canadian households (Canada, Statistics Canada 2000). In fishing-dependent households in our study, just under 50 percent of wives had paid employment. Both the nature of their husbands' work and the stages of the life cycle influenced these women's participation in the workforce. Approximately 95 per cent of all the women surveyed had at one time or another been engaged in the labour force. Most of these women left the workforce prior to or during the early stages of their marriages, but many returned to the labour market when their children went to school.[1] Table 7.1 summarizes the general characteristics of these women's employment. At the time of their interviews approximately 47 per cent of them had jobs. Of those currently working for pay, approximately two-thirds had year-round jobs, and the remaining third had seasonal employment. Nearly 62 per cent of deep-sea fishermen's wives who were employed had full-time jobs. They mostly worked day shifts (62.3%) or split shifts (34.8%) so that they could be home at nights with their families. Only 3 per cent worked night shifts. Coastal fishermen's wives who were employed had almost as many full-time jobs (56%) as their deep-sea counterparts. But of those who had jobs, most worked day shifts (69%), 18 per cent worked night shifts, and 13 per cent worked split shifts.

Table 7.1 Characteristics of Women's Employment by Sample (in percentages)

Characteristics	Coastal sample	Deep-sea sample
Ever employed	94.7	94.0
Employed in past but not now	47.7	48.0
Currently employed	47.0	46.0
Full-time – 35 hrs or more/week	56.1	61.8
Part-time – less than 35hrs/week	43.9	38.2
Daytime shift	69.0	62.3
Night shift	18.3	2.9
Split-shift	12.6	34.8
Employed at seasonal jobs	38.0	37.7
Current jobs		
Clerical	28.1	21.7
Service	43.7	40.6
Plant worker	5.6	13.0
Nursing/Caregiving	8.5	21.7
Teacher	2.8	1.5
Technical/Professional	2.8	1.5
Fisher	8.5	0

Missing values (0)

These women held jobs in a variety of areas. The service sector associated with retail, restaurant and other food services, hostelry, and other related jobs employed the largest number of women in our study: 44 per cent of coastal wives and 41 per cent of deep-sea wives. Many of these jobs remained seasonal, part-time, and paid minimum wage. Approximately 28 per cent of coastal wives and 22 per cent of deep-sea wives held clerical positions – specifically, secretary, bank teller, and bookkeeper. Deep-sea fishermen's wives participated more in plant work (13%), in nursing, and in other helping professions such as daycare work and social work (22%) than did coastal fishermen's wives (6% and 9%, respectively). Few women held teaching, technical, or other professional positions: 6 per cent of coastal and 3 per cent of deep-sea wives had these types of jobs. Nine per cent of coastal fishermen's wives worked for pay as fishers or baiters outside their own household fishing enterprise, but no deep-sea fishermen's wives did.

Tables 7.2 and 7.3 summarize the reported income distribution for wives with jobs, broken down by part-time and full-time employment sta-

Table 7.2 Frequency of Reported Income Distribution of Working Wives of Coastal Fishermen Broken Down by Employment Status (in percentages)

Wife's income median	All coastal $10,000–$19,999	Part-time less than $10,000	Full-time $20,000–$29,999
Less than $10,000	34.8	62.1	13.6
$10,000 – $19,999	33.3	34.5	32.4
$20,000 – $29,999	19.7	3.4	32.4
$30,000 – $39,999	4.5	0	8.1
$40,000 – $49,999	7.6	0	13.6

Missing values (0)

Table 7.3 Frequency of Reported Income Distribution of Working Wives of Deep-Sea Fishermen Broken Down by Employment Status (in percentages)

Wife's income median	All deep-sea $10,000–$19,999	Part-time less than $10,000	Full-time $10,000–$19,999
Less than $10,000	36.8	57.7	23.8
$10,000 – $19,999	39.7	30.8	45.2
$20,000 – $29,999	14.7	7.7	19.0
$30,000 – $39,999	2.9	3.8	2.4
$40,000 – $49,999	5.9	0	9.6

Missing values (0)

tus. After their husbands' incomes, wives' incomes made up the greatest component in household revenue. (Household finances are discussed in more detail in the following chapter.) In both relative and absolute terms, coastal fishermen's wives contributed more than did deep-sea fishermen's wives to their households' finances. The distribution of reported income varied between the two groups, largely on the basis of part-time versus full-time employment: the median reported for these incomes were less than $10,000 and $20,000–29,000, respectively.

Approximately one-third of coastal fishermen's wives reported making less than $10,000; another one-third reported making between $10,000 and $19,999; and 20 per cent reported making between $20,000 and $29,999. The remaining 12 per cent reported making $30,000 or more. This last group consisted only of professional women employed full time in teaching or in the helping professions. Approximately 37 per cent of deep-sea fishermen's wives reported making less than

Table 7.4 Frequency of Reported Income Distribution of Working Wives Broken Down by Educational Status for Deep-Sea and Coastal Fishermen's Wives (in percentages)

Wife's income	Deep-sea wife's educational level			Coastal wife's educational level***		
	Finished high school	Beyond high school	Total	Finished high school	Beyond high school	Total
Less than $10,000	43.3	32.6	36.8	51.4	13.8	34.8
$10,000 – $19,999	43.4	36.8	39.7	32.4	34.5	33.3
$20,000 – $29,999	10.0	18.4	14.7	13.5	27.6	19.7
$30,000 – $49,999	3.3	13.2	8.8	2.7	24.1	12.1
Total	44.1	55.9	100.0*	57.1	43.9	100.0**

Missing values *(1); **(6);
*** Pearson Chi Square value 14.399 with 3 degrees of freedom and a probability level of .002 (2-tailed)

$10,000; almost 40 per cent reported making between $10,000 and $19,999; 15 per cent reported making between $20,000 and $29,999. A little over 8 per cent reported making $30,000 or more. More deep-sea fishermen's wives were employed part-time, but they had fewer lucrative jobs than did comparable coastal fishermen's wives.

Differences in employment patterns – part-time versus full-time, seasonal versus year-round work, variations in pay scales – between deep-sea fishermen's wives and coastal fishermen's wives can be attributed to various factors. These factors include differences in levels of education and training, opportunities for employment, personal preference, availability of support services, demands of the household fishing enterprise, or deep-sea fishery, and the stages in the life cycle. When I broke down wives' reported income by education, I found a high correlation between education/training, and reported income for the coastal sample but not for the deep-sea sample (see Table 7.4). This difference between samples appears to be linked to employment opportunities, personal preference, financial need, and support services.

Economic and social opportunities for fishermen's wives partially depended on residency. The industrial deep-sea fishery operated out of a few large communities. Port facilities associated with modern industrial processing plants existed in centres with adequate labour for the needs of both plants and vessels. Usually, other businesses in the community offered services to the fishing industries and their employees, but they also supplied similar services to related industries. Larger communities also had major social service facilities such as schools, libraries, hospitals, transition houses, detox centres, and daycare centres, which supplied both jobs and support services. As a result, women living in these larger communities had a wider range of economic opportunities available to them, so personal preference for part-time versus full-time work and the types of work could be exercised. In smaller communities, where the economy primarily depended on coastal fishing, women found fewer economic opportunities outside of the fishery except in the local restaurant or motel. This scarcity of jobs meant that women from small communities who sought paid work, especially year-round professional employment, had to leave their home communities and travel to the larger centres. For coastal fishing-dependent households the additional expenses incurred through travel and from replacement of women's unpaid labour in the fishing enterprise with paid labour had to be traded off against the wife's earnings. Wives' income from paid employment had to make financial sense before these women were willing to enter the labour force. Moreover, these economic concerns were mediated by other factors: husbands liked having their wives work for them. Wives were more companionable, more malleable, and more willing to be self-exploitative than were hired helpers, who demanded formal compensation, insisted on specific hours of work, and did not put the fishing enterprise ahead of their own needs.

The characteristics of their husbands' work made it difficult for fishermen's wives to get and keep a full-time job. Among women who had a job, about 28 per cent of coastal fishermen's wives and 51 per cent of the wives of deep-sea fishermen indicated that their husbands' work schedule affected their jobs. The most common difficulties for both groups centred on arranging care for their children and rearranging their work schedules to meet their husband's needs or, in the case of the coastal fishery, the needs of the fishery enterprise.

Jobs for women in industry and the service sectors usually involved shift work. Plant and hospital jobs usually comprised eight-hour shifts for five days with two days off, or twelve-hour shifts for four days with two or three days off. Service sector jobs in restaurants frequently had split shifts, from 10:00 to 2:00 and 4:00 to 8:00, or all evening shifts. Such job schedules did not mesh with either fishery's work cycle.

In the case of coastal fishermen, the degree to which the wife participated in the fishery related inversely to her ability to take part in the labour force. The demands for women's labour varied with the coastal fishery cycle. In fact, for some women their participation in the coastal fishery totally precluded any involvement in the work force throughout the fishery cycle. (Brenda and Bruce's household, discussed in chapters 2 and 4 illustrates this situation.) Some women, like Brenda, who participated fully in only one or two 'fisheries' (i.e., they went lobstering but had limited involvement in groundfishing) found that seasonal employment which corresponded to their 'off' time was more appropriate than taking a year-round job. Other women, who participated less in the fishing enterprise, wanted a job with enough flexibility to allow them to run errands or to do household fishing business when their husbands needed help. Part-time employment offered them the flexibility they needed, but full-time employment did not. The need for flexibility and physical closeness, which would allow for a quick response time when the need arose, precluded women from taking any job, however lucrative, that was far from home. Women who did not directly participate in the coastal fishery (i.e., helpers' wives, professional women) had greater employment opportunities.

The deep-sea fishery presented a different but related problem for working wives. Deep-sea fishermen expected their wives to be at home and available when they came ashore, and when a wife worked full time, this expectation could not be met. Part-time employment created a possible compromise, although this arrangement did not guarantee that a wife would be at home when her husband returned from the sea. Frequently when her husband had shore leave, a wife would trade shifts with another employee, call in sick, or use 'banked' overtime in lieu of salary. A few women who had full-time professional jobs, like Margaret, would occasionally take a day off when their husbands were in port. As Margaret explained: 'He would just like to have me home when he's home. He just said that. In fact I took a day off just last trip when he was home, on the first day, and he said, "This is so nice." We just sat by the pool in the afternoon and we chatted and talked, and the kids were in the pool; they had friends here. And it is nice, but it's because it's like a vacation, right? Vacations can't last forever. Anyway, he would just like to have me home when he's here.' Margaret saw taking a day from work to be with her husband as a vacation. Another woman, a teacher, saw it as playing hookey from school. But the bottom line remained that neither of these women would give up their job or their independence just to 'be there for him.'

Connelly and Macdonald (1985: 416) argued that this need 'to be on call' accounts for the 'persistently lower labour force participation rates' of all fishermen's spouses, compared to fish plant workers' wives. Few jobs had, or allowed, the flexibility needed to respond at home, on short notice, as fishermen's wives required. A part-time job, when available, provided only a partial solution, but it did produce some of the benefits of employment.

Childcare and accommodating children's needs also limited the participation of fishermen's wives in the workforce. Some women found these demands, combined with the demands associated with their husbands' work, incompatible with having a job. But as we saw in chapter 4, having a family also limited women's job possibilities. Some work schedules were incompatible with childcare needs and the resources available to meet these needs. In our sample, those deep-sea fishermen's wives who worked night shifts had only adult children at home. None had dependent children. Coastal fishermen's wives who worked night shifts had to rely on their husbands or other family members to look after their dependent children. Women with school-aged children tried to schedule their jobs to coincide with school hours, and used after- and before-school daycare programs. But the women with children still in the home generally called upon family members, including some husbands, to help out with childcare. But, as we discussed in the previous chapters, husbands seldom became the primary childcare worker when they came ashore.

In order to balance all of their activities, these mothers had to have a dependable support network and had to be extremely well organized. As Gail explained:

> I'm involved in a lot of volunteer organizations, as well, with the fisheries exhibition, with the nursery school that my daughter attends, and that. So it means that you can't have disorganization in your household. [They] have to eat meals with me. It's a very strict schedule as to when we eat, when bath time is, when bedtime is, when music practice is, because you have to keep everything going right. And basically I find that when Gordon is out, from the time I get home from work until the time I go to bed, I don't get a chance to just sit down and leisurely do a craft or something another. You're constantly on the go because there's so much more to be done when it's only one. And, basically, you are a single parent family during that time frame.

For both coastal and deep-sea fishermen's wives, their husbands' jobs defined and confined their employment. Although each fishery puts different types of pressure on their households, the requirements for coping with these challenges – flexibility, organization, and support networks – were the same. For many women, just keeping it all together – home, job, social life, marriage – was difficult at times. Gilda, a coastal fisherman's wife, had full-time employment and also worked as the shore manager and bookkeeper for the household fishing enterprise.

> Life's too hectic. I've got too much to do to really socialize. Like if you have any spare time you've gotta try and get the house cleaned up and your bookwork done, because the bills come. I file them behind the radio, then I pay them and I file them in the filing cabinet. You know, it's too much to do. I don't have any time, I don't know why. You're just busy, busy, busy, right? You just get up in the morning and you just zoom, zoom, zoom, zoom, zoom, all the time. You just throw a forty-hour week in there that you're not home. Sweep the dirt under the mat until the next day you vacuum it up (laughter). But no, life's very high-paced. People say they're bored, I don't know why they're bored.

For deep-sea fishermen's wives their husbands' presence in the home added another level of interaction. But for both coastal and deep-sea fishermen's wives this juggling act took its toll. Women spoke of being tired and irritable, and of having to organize everything 'just to get it all in.' As Margaret explained: 'I guess I'm tired. I guess by the time that I'm finished my work, and I'm finished being a mother, right? There's all my extra activities I have to prepare and plan, there's not a lot of time [left for him or me].'

The Life Cycle and Employment Histories

The employment history of fishermen's wives was anomalous in the North American context. It still related directly to the life cycle. Consider Holly's employment history: 'I worked until we had our family, and then [there were] twelve or fourteen years that I didn't work. Then I went back to work and work[ed] off and on for the last few years, say probably six or eight years. But the last two and a half years it's part-time and casual, and, like I said, sometimes it involves more than eighty hours into a week or into two weeks. So I'm able to do that now that the family is older. It worked out pretty good. Basically, I guess we were lucky. It was day shifts at that time, and my worked seemed to go along hand-in-hand with his fishing. So it wasn't a great problem.' Holly's employment history again illustrates how many women attempted to juggle their needs, the needs of their husbands and children, and the financial needs of their household.

Holly's experience represented a common pattern in the surveyed households. When the women in our study married, they either had just left school or were participating in the labour force. I asked those women who had been employed in the past but had dropped out of the workforce sometime during their marriage, approximately 94 per cent of our sample, why they had left their jobs. In particular, I asked, 'What specific event marked your leaving your job?' (Table 7.5 summarizes the data.) For both coastal and deep-sea wives, marriage and setting up their current household represented the most common reason for stopping paid work. Almost 70 per cent of deep-sea fishermen's wives left their jobs when they got married; 53 per cent of coastal fishermen's wives did. The second most common reason was starting a family – pregnancy, the birth of the first child, and the births of subsequent children. Approximately 17 per cent of these deep-sea fishermen's wives and 19 per cent of these coastal fishermen's wives stopped working for this reason. All of these events occurred relatively early in marriage and were related to family/ household formation. (The difference in leaving patterns may be attributed to the relatively higher importance of the wife's income in coastal fishing-dependent households.)

Of those women who subsequently left the workforce, about 14 per cent of deep-sea fishermen's wives and 28 per cent of coastal fishermen's wives continued to work through the family/ household formation period. They later left the workforce for a number of reasons. Some of these women cited 'job-related reasons,' which included plant lay-offs and closures (particularly in the coastal sector), getting fed up with the job, illnesses, injuries, and retirement. Others noted family demands, such as the need to babysit grandchildren so that daughters and daughters-in law could go to work, or to look after infirm parents or parents-in-law. Still others cited illnesses and injuries to other family members, or pressure from their husbands to stay home. About 6 per cent of employed women never left the workforce – all were professionals.

Staying Home

In Nova Scotia fishing communities there is a social expectation that a woman will stay home and engage in full-time domestic labour when

Table 7.5 Event Marking Homemaker Status for Deep-Sea and Coastal Fishermen's Wives (in percentages)

Event	Coastal sample	Deep-sea sample
At marriage/begin common law relationship	53.2	69.1
Pregnant	1.2	1.2
At birth of first child	12.7	14.9
Child born	5.0	1.2
Injury/illnesses	1.2	3.7
Laid off / seasonal work	11.4	4.9
Retired / quit job	6.4	2.5
Family issues*	8.9	2.5
Number of cases	79	81

Missing values (0) *This category includes issues such as birth of grandchild, husband's displeasure in her working, the need to look after parents.

her children are very young. But staying at home did not come naturally to many of these women. As their children grew older, the wives' desire to return to the workforce usually increased. For example, at the time of her interview Donna had decided to return to work:

> She'll [her daughter] be a year on Sunday. I wouldn't go back to Halifax [to work], I'd be just gone too long through the day. But I'd like to get something around here – part-time, full-time – whatever it takes me (laughter). Anything to just give me some pocket money (laughter). That would just be it. Do I take her to a daycare? I'd probably be paying out more in a week than I'd be getting back from working, but to him it's, like, ridiculous, right? Why would you go and pay somebody more money than you'd be making in a week. But to me, I'd still be, 'well, I'm working, right?' If I said, 'I found a job, I'm going to work.' But we've discussed it a million times and the final outlook was, he said, 'Look, if you really have to work, if you think you have to do it – but I would prefer you to stay home,' type of thing.
>
> Well, he is so understanding about it [that] it kind of annoys me, right? If he'd say, 'No, you're not working.' Well, [I'd say] 'Yes, I am.' Then I would work, right (laughter)! But I can see his point, you know, having somebody else look after her, and who would I get, and that would be half the battle. If I had somebody that I trusted with her, I would probably go tomorrow, type of thing, and start beating the pavement looking for a job. That's probably the main issue ... who would take care of her?

Employed mothers worried about the need and quality of childcare. For Donna, childcare concerns had two components: first, who would look after her daughter; and, second, would there be a net financial gain once her childcare costs were subtracted from her pay.

Although many women talk about going back into the workforce, few made this change until their children enrolled in school. Even when their children attended school, mothers still worried about their care after school, at lunchtime, or when they become ill.[2] For these reasons, some mothers talked about going back to paid-work only when all their children had reached their teens. Frances's situation illustrates these concerns:

> I would like to be working to try to help with the bills, but when the kids are older. I know Fred Jr, he's thirteen. He could look after Frieda and Frank, but they wouldn't listen to him, so it's no good there. And when you think about Fred working and me working, it would be senseless for me to work anyhow, because what money I make working would all go to income tax. So it doesn't make sense to do it. Fred doesn't want me to work. He wants me to stay home. I would love to do something to help contribute. I enjoy looking after the kids now. See, I'd like to help financially because, well, we want to have our own house instead of renting.

But Frances felt that it was still too early for her to enter the workforce. Although all their children attended school, she did not feel confident about leaving them alone. She also questioned the financial benefits if she were employed. Her husband's desire that she stay home increased her reluctance to enter the labour force. At the time of her interview Frances did not yet feel compelled, financially, to enter the labour market. After the children reached their early teens, women did more readily enter the workforce. Holly went to work when all her children were teenagers. She described them as 'latchkey kids.' As she explained: '[I didn't go to work], not while the children were growing up. That's wrong! When they were probably twelve and fourteen, I went back to work. When they were able to come home and let themselves into the house and be reasonably responsible until I got home, [then] I went back to work.'

However, these were systemic problems. For women with young children who continued working, or for women who only stopped working to take maternity leave, or for women who wanted to return to the workplace, finding someone to care for their children dominated their con-

Table 7.6 Husband's Support for Wife's Work Status Reported by Deep-Sea and Coastal Fishermen's Wives (in percentages)

Husband's opinion	Coastal sample			Deep-sea sample		
	Wife not working	Wife working	Total	Wife not working	Wife working	Total
To have a paid job	11.4	26.8	18.7	11.3	33.8	21.6
To stay at home	64.6	29.6	48.0	68.8	26.5	49.3
Doesn't matter	24.1	43.7	33.3	20.0	39.7	29.1
Number of cases	79	71*	150*	80*	68*	148**

Missing values *(1) **(2)

cerns. In many fishing communities formal daycare was unavailable, only operated seasonally, or had restricted hours of operation. The few year-round daycare facilities in the area were overcrowded, expensive, and tailored to nine-to-five, Monday through Friday jobs. Many women I interviewed felt that putting their children in daycare indicated their neglect of them, because many mothers perceived daycare centres as typically underfunded, understaffed, and under-equipped. As an alternative, women channelled their energies into informal, individual childcare solutions that at least permitted them to think of themselves as good mothers. But this solution perpetuated the problem of inadequate formal daycare facilities appropriate to fishing families.

Another frequent obstacle keeping women from the workforce was their husbands' disapproval of it while their children were still young. In the survey I asked women if their husbands preferred that they be employed, or stay at home, or if he had no preference. Table 7.6 summarizes these data. Nearly half of all women surveyed said their husbands would prefer that they stayed home, while approximately 20 per cent reported that their husbands would like them to have a job. Only 11 per cent of women currently unemployed said that their husbands would like them to have a job. Nearly 30 per cent of coastal and 27 per cent of deep-sea wives currently in the labour force said their husbands would prefer that they not be employed.

As these numbers show, about half of all husbands, and between 25 and 30 per cent of husbands with employed wives, did not want their wives in the workforce. Husbands' desire to have their wives stay home with young children reflected a belief that only the mother could provide proper care for their children. Even when the children attended school, some husbands still wanted their wives at home. Some husbands also saw their wives' going out to work as a sign of their failure to support their family adequately. These attitudes sometimes also meant a desire on the part of the husband to have his wife dependent on him. Yet some wives, like Donna, whom we discussed earlier in this chapter, defied their husbands and got a job. As Karen, another assertive wife, explained: 'We're an independent lot (laughter). I worked when we were married. I quit when I got pregnant with my oldest child, and I always said if I had kids I was gonna be home till they're going to school. He goes to school in September, I started a job two weeks ago, and I'm not about to sit here. No, I just can't do it. I've missed work. I'm glad to be back. And he understands. I mean, he's not sitting complaining about [it], but he said, "Well, this is going to be different," and "I'll always look at it as a necessary evil," and all that stuff (laughter).'

For both deep-sea and coastal fishermen's wives, community norms discouraged mothers from going back to work when they had young children. Women struggled to do what appeared to be best for their families and themselves. They talked about trading off an enhanced financial situation for the household, a good situation for themselves, and the best for their children. The lack of formal daycare facilities and other services made the choice more difficult. So why did they go back to work?

Although our survey indicated that most women quit their jobs during the period of family formation, about half of these women eventually returned to the workforce. About two-thirds of both coastal wives and deep-sea wives, currently employed, had an interim outside the labour market. What led to their re-entry to the labour market?

The simple answer was finances. Each stage of the life cycle exerted specific pressures on the household's finances. When couples married, they wanted to buy, furnish, and maintain a home. In the case of the coastal fishery households they also tried to save for or pay for and maintain a fishing boat and gear. These couples had little or no savings to draw on to cover these expenses. Such financial needs put pressure on young wives to retain or get a job, and once they had children their financial problems often worsened. As Joan, in my earlier 1987 study, put it: 'Then the finances. There were times, especially starting out when we were new[ly married], and we had so little money. You didn't know how you would manage. You'd rob Peter to pay Paul ... Of course, I was lucky. I'd started back to work. That was another thing when I first got married that people expected – she's going to stay home and bring up children. That was fine with me ... until you had a child and you realized that you can't make it on just one income, not if you wanted a home ... Money can be a real problem. Some of those women alone that aren't working, I don't know financially how they did it' (Binkley 1994: 78–9).

Women employed in the middle years of the life cycle, when their children attended elementary school, junior high, and high school, said they worked for necessities, for 'extras,' or for an improved lifestyle for their families. Here's how Holly looked back on those years: 'We were never in want, and the children were never in want, but if it was some trivial thing, well you'd say, "Well it's my money. I'll get it for them," or this type of thing. You would think differently because it's your money and you've earned it, and I guess you think, well if I want to do that I'll do it. And he'd never stand in my way and say you shouldn't have done it, or whatever. He just let me do what I wanted to do with it. And I used to buy things for the house that maybe I normally wouldn't have done, had it been his pay or his money. You'd just sort of say, well. So that's what I'd do.'

In the later years of the life cycle new expenses emerged, such as saving for retirement or for costs associated with caring for parents or grandparents, or for sending children to university. Moreover, in the later years, as they approached retirement, fishermen began to cut back on their time at sea. As the gruelling working conditions became harder to sustain, men began taking fewer trips and/or shorter trips each year. With the subsequent decline in the fishermen's income, the wivès' jobs took on additional importance. Some women entered the workforce at this time; others remained employed – although many would have liked to give up their jobs – because they felt their paycheques were no longer just for 'extras,' but had become essential to the survival of the household. Elizabeth's situation illustrates this point:

> Well, I do enjoy working. I think my job now is more necessary than when I first went to work. When I went to work, I went because I felt that I wanted to get out and work and meet more people and do that sort of thing. That was true for many years. But I think now, financially, with three children just at that age of university and so forth, that it's a necessity, and where Eric's job, too, has been cut. He's not earning the same salary he was, say,

148   Set Adrift

four years ago. So I feel now it's a necessity, which puts a different light on my job – I liked it better when I didn't actually have to do it. It was always like the choice was always there. I think if I really wanted to retire, I don't think I could for several years.

Women asked themselves: What is best for the children? What is best for the household? What is best for me? As households moved through the life cycle, demand for enhanced financial support grew. As children aged, mothers' perceived need to stay home with their children lessened. As the balance between these needs changed, women entered, re-entered, or dropped out of the workforce.

Who's Working Now?

We now turn from looking at the structural relationship between employment and the life cycle to considering individual reasons for returning to the workforce. For those women who returned to the labour force, the decision to get a job was an informed choice. They had experience that would allow them to anticipate the challenges and the advantages that employment would bring.

Their reasons for being employed often related to the qualities they liked most about their previous job. In the survey, I asked women currently employed to choose the most important qualities of their job from a list of twelve characteristics of paid work. I also asked them to cite additional qualities of their job that they found important. Tables 7.7 and 7.8 summarize these results, and present them in ranked order for the coastal sample (column 1), with the deep-sea sample (column 2) forming the comparison.[3]

When we look at the ordering of the most important qualities of paid work, 'take home pay' ranked number one for both groups of women – over 39 per cent of coastal and 37 per cent of deep-sea fishermen's wives (see Table 7.7). These women ranked 'being able to do my own work' (no hassle) and 'enjoy doing my work' as the next important factors, although in the opposite order of importance in each group.

In Table 7.8, I aggregated the data regarding the three most important qualities of one's job. In this case, it did not matter which order the respondent cited the quality of work because I gave each response the same weight. For example, if a woman cited take-home pay, benefits, and getting along with co-workers, in that order, I gave her response the same score as the woman who responded getting along with co-workers, bene-

Table 7.7 Most Important Quality of Their Job for Deep-Sea and Coastal Fishermen's Wives by Sample Ranked According to Coastal Sample (in percentages)

Reason for being employed	Coastal sample	Deep-sea sample
Basic take-home pay	39.4	37.7
Being able to do my own work	14.1	14.5
Enjoy doing my work	14.1	21.7
Benefits	5.6	2.9
Getting along with co-workers	4.2	7.2
Service	5.6	2.9
Job security	2.8	4.3
Total hours of work	5.6	0
Getting along with company management	4.2	0
Opportunity for promotion	2.8	1.4
Shift schedule	1.4	4.3
Health and safety	0	2.9
Number of cases	71	69

Missing values (0)

fits, and take-home pay. The results, then, represent the frequency with which these women chose some qualities over others. The results show different patterns for the two samples, and reflect the differing demands and characteristics of the two types of household (see Table 7.8).

In the survey I asked women, in an open-ended format: 'What are your three most important reasons for going to work?' Their responses can be summarized under four categories: monetary, social, psychological, and work-related. The most frequent reason cited for taking employment was monetary. Reasons for this ranged from the need for more money for themselves or the household, or savings such as registered retirement plans, to the need for drug, dental, or health care plans for themselves and their families. The social and psychological reasons for working included factors such as the need to get out of the house, to avoid boredom, to retain their mental health, and to keep busy. Meeting new people and talking to adults also appeared frequently as additional reasons to be employed, as did increased independence and personal growth. In only a few cases did attributes of the job, such as the enjoyment of the work itself, the honing of skills, and/or the development of additional skills, appear to be important.

For most of the women in our study, working meant paying the bills and getting out of the house when the kids were attending school. It did not mean having a career. Many women who had a job simply liked hav-

Table 7.8 Compilation of Three Most Important Qualities of Their Job for Deep-Sea and Coastal Fishermen's Wives by Sample Ranked According to Coastal Sample (in percentages)

Reason for working	Coastal sample	Deep-sea sample
Enjoy doing my work	59.2	58.0
Basic take-home pay	57.7	65.2
Being able to do my own work	31.0	37.7
Benefits	29.6	17.4
Getting along with co-workers	19.7	39.1
Service	18.3	10.1
Job security	17.9	8.7
Total hours of work	15.5	13.0
Getting along with company management	14.1	13.0
Opportunity for promotion	12.7	4.3
Shift schedule	9.9	10.1
Health and safety	5.6	5.8
Number of cases	71	69

Missing values (0)

ing the extra money to spend. As Margaret said, 'I like the money a lot. I like the things that money can buy.' Others complained about the boredom of everyday life with small children and limited adult company. They talked about going back to the workforce to protect their sanity. Employment created as well a social environment for making new friends and enhancing leisure activities (Deem 1986; Gloor 1992; and Wellman 1985). Once women returned to the workforce they satisfied many of these needs for adult company, having 'her own money,' and relief from the tedium of being at home. As Karen stated: 'It's wonderful [to be back at work]! It's good to talk to people at eye level for a change, instead of looking down (laughter). But I'm enjoying it a lot. Oh, as far as the money part of it goes, it's probably not going to amount to much, but I'm out of the house.'

For others, like Opal, personal growth was the most important aspect of being employed. As she explained, her job helped her to become more independent and self-reliant, skills she needed to cope as a deep-sea fisherman's wife: 'The job was very important because I got to meet people. And, I loved the people. Not all of them, but you know most of them. Like with him out in the boat too, it gave me an outlet and it got me over my shyness with people, and then when he's home, like he's doing his thing. We talk about the kids and fishing, but we haven't got a lot other things to communicate about. So this was good, because I had all this variety (laughter). I talked about everything and anything with people, and met so many interesting people.'

Of course, for a few professional women, like Margaret, who had been employed all their adult lives, working outside the home was a career choice, a way of life. As Margaret put it: 'Oh it's everything. I like the social aspect of it. I work with a great bunch of people. I have a good job and I guess there's some prestige that comes with it, and I like that. I like helping people. And I really like the independence. Like, I've never not worked, and that scares the pants off me – the thought of not having my own money, because I've never not worked.'

The growing independence that working wives developed altered the power dynamics in the household. Sarah and Stan had been married for almost ten years before they separated. Although Sarah felt stifled in her marriage, her emotional and psychological dependency on Stan, a deep-sea fisherman, kept her from leaving. However, once she got a job in a bakery, became financially independent, and started making new friends, she no longer needed to depend on Stan for financial or emotional support. As Sarah explained: 'I went to work and I started making other friends. Then when the opportunity came along for me to manage the bakery, well that just opened up a whole new horizon for me. Stan said to me not long ago, "The whole problem with you started when you started running that store and got that independence. You didn't need me anymore." And I suppose he was right.' For Sarah, her job began as a way to get out of the house and ended as a way to dissolve an unhappy marriage.

In the interviews it became apparent that the current fisheries crisis had affected women's attitude towards their own employment. Employed women felt that their present job had taken on greater importance. Many women said they no longer worked to supplement their husbands' wages; rather their own wages had become the crucial ones. Older women spoke of staying or continuing to work rather than retiring as they had planned. Women who were unemployed talked about getting a job.

In the survey I asked those women who were without paid employment if they were looking for paid work, or if they had plans to look for a job in the coming year. Nearly 14 per cent of coastal wives and 10 per cent of deep-sea wives said that they were actively looking for a job. An additional 58 per cent of coastal and 43 per cent of deep-sea thenunemployed wives planned to start looking for employment over the next year. The type of employment they were considering included jobs in the clerical, sales, or service sectors.

Because 'looking' and 'planning to look' for employment differed substantially from actually having a job, the constant financial worries associated with the declining fishing industry dominated most of the discussions surrounding women's plans for the future. These women cited the impact of the fisheries crisis on their family income as the most common reason for returning to the workforce. The greater number of coastal wives looking for, or planning to look for, employment reflected the greater financial strain on the coastal fishery. Even those women who said they were not planning to look for a job talked about the inevitability of their joining the workforce to help support the family if the fisheries crisis continued.

Anne, a coastal fisherman's wife with three children, felt that she would ultimately have to get a job.

> Well, we talk about it. When he was making a lot of money, 'Oh no, it's perfect that you're at home. I don't have to worry about the kids.' When things got a little tight, it was, 'Well why don't you go get a job?' And it's almost a slap in the face because for years it's okay that you've stayed home and had babies, and then it's like you're not contributing anything. It's like there's been a twist somehow. And we've actually had major arguments over that. And now he does see it my way, like, 'You're right. The day will come when you'll go out to work,' and he just has to plug along a little harder till I don't have to leave my kids with a babysitter.

Although Anne had no intention of getting a job until their children were older, she had started preparing herself to return to the workforce. Before she married, Anne worked in an office, but she realized her skills had deteriorated since then. 'I skim through the newspaper every day. A lot of [jobs] require training, like the computer and everything. It's almost scary for me to think about going back to work, because I've been out of the workforce for, it's been twelve years.' In order to upgrade her skills, she enrolled in a computer class at the local high school.

Some women saw retraining as an opportunity to get a good job. However, there were two recurrent problems. First, the retraining programs offered locally did not lead to lucrative job opportunities. Costly retraining programs, such as advanced computer skills, existed only in the Halifax area, at least an hour's drive away. Second, once women received training in these skills, few such jobs existed locally.

I Want My Own Money

Many women expressed a concern about dependence on their husbands' income and said it was a major reason for returning to the workforce. Donna's description of her situation revealed her frustration at being home alone with a small child and not having her own money: 'Well, he wants me home because he doesn't want somebody else taking care of the baby. And I want out because I want to socialize (laughter). You know, doing the baby talk thing all day, and I like having my own money, even if it's just ten dollars, it's my ten dollars (laughter). And, you know, asking him for money for gas and stuff like that [I] can't handle it.'

Whether wives were employed or not, the relative value of a wife's contribution to the household through domestic labour, compared to the husband's monetary contribution, remained an issue. A wife had to develop a sense that her domestic labour had a monetary equivalent; her husband had to learn the real value of her domestic labour. Once she recognized her work in the home as 'real labour' – that is, it had a monetary value – then she appeared to be able to take 'his' money without feeling financially subordinate.

In other households, whether the wife was employed in the labour force or not, fights about 'our' money versus 'his' money stemmed from his not accepting her domestic labour as a real contribution. As Brenda, a mother of three children, with a strong sense of her contribution to her household, stated:

> I don't have any pay anyway, so his money is my money, and what I make on baby bonus is basically his money. Like there's no his, mine. I mean, in heated arguments maybe there's, 'Well, you haven't contributed anything to the relationship.' But you have to say, 'I have so! I nursed three babies; I must've saved a couple thousand dollars on Similac [breast milk substitute] right there.' Like, in my own way I have contributed a lot. And if he told me otherwise, it would bother me, probably hurt me. It can come up and it wouldn't matter because I know I have contributed enough. So money, it isn't a cause for a lot of arguments because you have kids involved, but for the most part we try to avoid the topic.

As we saw earlier, Brenda not only contributed her domestic labour she also worked as the shore manager for her husband during lobster season. Although money did not 'change hands,' she saw herself as being 'employed' by him for six months of the year. But the question of whose money it was continued to be an issue. Brenda gave herself away when she said, 'We try to avoid the topic.' By avoiding the topic, I knew it remained an issue below the surface, an issue which came into play in the heat of an argument.

Not just newly married women and young mothers felt this way. Women who did not quit the workforce during the period of family formation, but gave up their jobs later in the life cycle, also voiced these concerns. These women found that being supported solely by their husbands created tension between them. For these women, their need for their own money related to their identity and independence, especially if they had been professionals such as nurses or teachers. However, some women did have their 'own money' – pensions, investments, or other forms of income – which mitigated their feelings of lost independence and self-esteem. Just having their own 'pocket money' meant that they did not have to depend on their husband's money.

When the discussion of finances came up in the interviews, women frequently spoke of 'our' money, 'his' money, and 'my' money. By analysing this discourse we can identify discrepancies between how people talk about their financial resources and dispersions, and their actions. We can also see how money relates to power in the household, even though the rhetoric of the household is one of equality and joint participation.

Let's look at two examples, which we will return to in the next chapter. In the first case, Tim, a coastal fisherman, and his wife Trudy, who did not have paid employment, said they set priorities jointly for major household purchases, but Trudy deferred to Tim when it came to the needs of the fishing enterprise. In her interview Trudy described this bargaining session. They had agreed to a number of items related to the fishing enterprise, and moved on to decisions concerning household items. Tim wanted a new chesterfield set; Trudy wanted a CD player. Trudy deferred to Tim once again. In the second example, Donna, formerly employed, was staying at home and looking after their child. Donna and her husband, Dean, a deep-sea fisherman, used to contribute to and manage their household's finances jointly, but since Donna quit work Dean has controlled the finances. In both these cases the final argument used to justify actions and satisfy the interviewer was, 'Well, it's his money.' Thus for Trudy and Donna the ability to generate income for the household translated into the ability to control how household money would be spent. In other words, the value of their domestic labour did not equal that of their husbands' labour.

Let's look at another of Donna's statements: 'It's not mine. It's just a mental thing, right? Knowing that I'm spending his money, and it just, I don't know, it just bothers me.' I kept coming back to the distinction between 'his' money and 'our' money. I asked if Dean asserted that it was only his money to spend, and if he restricted her spending. She replied, 'No, not with his money, that's for sure. He keeps trying to beat it into my head that it's *our* money, but it's not *our* money. I've got myself convinced, he made it, it's his money, it's not my money. And then he says, "Well, you're taking care of the baby, you're keeping the house," but it seems so minor. Well it is hard work, but I don't know. I guess I'm kind of old-fashioned – [I] think it's, like, my job to, it's just something that should be expected of me, you know what I mean?' (emphasis Donna's). From Donna's statement it appeared that she had not yet accepted her domestic labour as equal to Dean's contribution, although he had done so, rhetorically. Remember that Dean controlled the finances and Donna had just decided to return to work.

The question of whose money it was and how it should be spent appeared particularly difficult when the wife had just left her job. As Linda explained: 'And then the hardest part is when you gave up your job, your unemployment ran out, and then you have to start spending his money. Like each time he came home I would say, "I spent it here and here and here." You were trying to keep lists of where the money went because you felt guilty for it.'

These women's comments exposed their feelings of guilt that they no longer contributed financially and therefore had lost their entitlement to spend household money as they chose. As Donna explained: 'Then [when I was working], my money was my money, and I didn't touch his money when I was working. So it was really hard for me. Like all his money was all his money, type of deal. And I paid little bills like the phone bill, but he would get the bigger bills, like power bills and stuff like that.' The transition was exacerbated in Donna's case because she and her husband moved from a pooled management system where they jointly contributed, managed, and controlled household finances to a system where only he contributed to the household finances but she managed the money. In this new system Donna felt she was not, and could not be, an equal participant in the decision-making process.

But there was a flip side to this story. When women joined the workforce, they spoke about newly found independence. As Opal commented, 'I was really independent. I didn't have to ask him for money, I had my own. If I wanted something, I went and got it.' On the other hand, Holly went to work

> to get out [of the house]. To do something on my own and have some spending money of my own – not that I didn't have any of my own, but it's different. When you have your own paycheque, it's a great independence. It does a great thing for you. You just think, this is mine I can almost do with it as I wish. Because my husband is the type of person, he's not a selfish person, if it wasn't to pay the bills and whatever, he would just say, 'It's yours, you've worked for it. You take it and do with it as you want.' And I guess that's probably what I did when I wasn't really obligated, when I first started working. I was getting out, and I had spending money, and if I wanted to buy something extra for the children, I bought it, and I didn't feel quite so guilty.

By having a job, Holly increased her self-esteem, developed new interests, and brought home a paycheque that could be used to buy things that she wanted for herself or for the household. But it went beyond just having more money. For Holly, having her own money differed from having access to money before she had a job. Throughout this process, she realized the accommodation that she had made in accepting her husband's money. Having her own money had freed her from that accommodation and from the guilt associated with taking his money.

What women did with their money depended on the financial situation of their households.[4] In some households the wife's money – considered 'fun money' – purchased such things as clothes, make-up, treats for her children and husband, meals outside the home, or travel. In others, the wife's money – considered 'essential money' – purchased such things as basic household supplies and groceries, children's clothing, and other items that her husband's income could no longer cover. In some cases, where women originally saw their paycheque as merely a supplement to the household finances, their financial contribution had become essential to the household. For example, Holly originally got a job for the non-monetary benefits – independence and an enhanced social life – but since Holly's husband had been laid off, her income had to sustain the household.

As we have seen, once women became contributors to the household finances, they felt they had more of a say about how it was spent. They saw themselves not merely as managers of the household funds but as financial partners in the fishing enterprise. In many cases this meant more than not feeling guilty when they spent money on less essential purchases: it was a way to voice their preferences concerning how to spend the household money. But when women stopped working they felt that any real contribution to the household had ceased. In these circumstances a woman's self-esteem depended on both spouses' recognizing that her domestic labour had sound economic value.

In this chapter we saw how several fishermen's wives viewed the advantages and disadvantages of employment for themselves and their households. We also examined how the internal dynamics in the household changed when these women took on or gave up paid work. In the next chapter we will continue to explore these issues, and we will look as well at how financial resources were managed and used, and how such strategies, in turn, reflected household dynamics.

CHAPTER EIGHT

Our Money, Your Money, My Money

I think this holds true for a lot of fishermen's wives. It goes one of two ways. Either the fisherman's wife knows absolutely nothing about her husband's income and doesn't look after anything, he does it all, or it's completely the opposite and the fisherman earns the money and the wife looks after everything, and the fisherman never sees where any of it goes. Well I was in a situation where my husband earned it, and I looked after everything. I had responsibility for everything. (Sara)

Every fishing endeavour assumes that all players – captains, crews, helpers, and owners – take the risks and either reap the benefits or share the losses. In the coastal fishery, the captain/owner paid the crew on a 'share' basis. Each fisherman or 'shareman' received a specific portion, or share, of the profits of the catch after expenses had been deducted from these proceeds. Expenses included all joint costs such as food, fuel for the boat, or labour for baiting trawl. The boat's share, usually 25 per cent of the profits, covered the capital costs related to the boat – mortgage and loan payments – and gear (including repairs). The amount of labour a person contributed both on- and offshore determined his share. Moreover, the share could be increased by the contribution of a spouse's labour onshore or at sea. For example, a crewman's wife could bait trawl to lower her husband's expenses so that he would receive a greater share of the profits.

In the deep-sea fishery the share system had been modified and bureaucratized. Prices for specific types of fish and levels of quality, and job classifications as set out in collective agreements, determined the share individuals received. Companies told captains where to fish and how much of each type of fish of a particular quality they wanted caught and delivered to a certain port on a specific day. As companies increased their ability to control operations from shore – through better communications between the vessels and the head office – they instructed captains according to the needs of plants and markets. Companies rewarded the most lucrative trips to captains who met their quotas and consistently delivered their catches on time. Under this system, with earnings for workers depending on catch size and fish type, fishermen's incomes had become more stable and uniform. Fishermen could increase their earnings only by catching particularly high-valued species, by getting bonuses for quality, and by sailing with captains favoured by the company.

However, the specific characteristics of the fisheries, whether deep-sea or coastal, still made predictions of actual earnings difficult. Catching fish, no matter how technologically sophisticated, remained hunting a resource at sea – a specific catch cannot be guaranteed. Moreover, the fishery had a seasonal rhythm. Fishermen's wages, whether in the coastal or deep-sea fleets, fluctuated throughout the year. Nothing was predictable – the catch, the weather, or the reliability of the fishing technology. Not until the catch came ashore did the fisherman know the amount of his pay. As Amy explained: 'You know once a month what you're going to have. It all depends on where they're going, but normally Georges, Browns, or Germans [fishing banks] is a decent paycheque. It's just scalloping: one day you're rich, the next day you could be poor. You don't know with any fishing.' This inherent characteristic of the fishery meant households needed to be financially flexible in order to meet their future needs.

Deep-sea fishermen made substantially more money than coastal fishers. (See Tables 8.1 and 8.2 for fishers' reported income distribution.) In 1992, 10 per cent of deep-sea fishermen reported making less than $30,000 and 10 per cent reported making more than $70,000, with the median reported income falling between $50,000 and $60,000. A quarter of coastal fishers reported making less than $20,000 and 10 per cent reported making more than $40,000, with the median reported income falling between $20,000 and $30,000. Furthermore when we break down the samples by job status – captains/owners and helpers for the coastal sample, and captains, other officers, and crew for the deep-sea sample – then even greater differences emerge. As indicated in Table 8.1, helpers on coastal vessels made a median reported wage of $20,000 to $29,999 annually. Only 12 per cent of this group reported making the top wages

Table 8.1 Frequency of Reported Income Distributions of Coastal Fishermen Broken Down by Job Status (in percentages)

Husband's income Median	All coastal $20,000–$29,999	Captains/owners $30,000–$39,999	All others $20,000–$29,999
Less than $20,000	24.6	24.7	24.2
$20,000 – $29,999	26.9	24.7	33.3
$30,000 – $39,999	30.0	29.9	30.3
$40,000 – $49,999	8.5	7.2	12.2
$50,000 – $59,999	4.6	6.2	0
$60,000 – $69,999	3.1	4.1	0
$70,000 – $79,999	0.8	1.0	0
$80,000 or more	1.5	2.1	0

Missing values (0)

Table 8.2 Frequency of Reported Income Distributions of Deep-Sea Fishermen Broken Down by Job Status (in percentages)

Husband's income Median	All deep-sea $50,000– $59,999	Captains $100,000 and over	Other officers $40,000– $49,999	Crew $40,000– $49,999
Less than $20,000	2.8	0	4.3	2.4
$20,000 – $29,999	7.6	0	8.7	8.3
$30,000 – $39,999	16.6	6.7	8.7	22.6
$40,000 – $49,999	21.4	0	30.4	20.2
$50,000 – $59,999	20.7	0	13.0	28.6
$60,000 – $69,999	13.1	20.0	13.0	11.9
$70,000 – $79,999	5.5	0	10.9	3.6
$80,000 or more	12.5	73.3	10.9	2.4

Missing values (0)

of $40,000–$49,999. Although coastal captains had a median reported income of $30,000 to $39,999, over 13 per cent reported making $50,000 or more. One man reported making over $100,000. But as Table 8.2 indicates, the real money was made in the deep-sea fishery. Here over 70 per cent of the captains reported making $80,000 or more and had a median reported salary of over $100,000. The outliers reflected the impact of the fisheries crisis. One outlier, an older man, had been employed part-time as a relieving captain and had subsequently been laid off. The others included captains who sailed on double-crewed vessels. Other officers and crew reported a median

Table 8.3 Sources of Husband's Reported Income for Fishing-Dependent Households Broken Down by Deep-Sea and Coastal Fishermen (in percentages)

Sources of income	All fishing households		Coastal households		Deep-sea households	
	N	%	N	%	N	%
Wages only	46	15.3	15	9.9	31	20.6
Wages + UI	241	80.1	132	87.4	109	72.6
Wages + UI + WC	12	4.0	2	1.3	10	6.6
TAGS	2	0.6	2	1.3	0	0
Sample size	301		151		150	

Missing Values (0)

income of $40,000 to $49,999. Once again lower incomes reflected part-time, or junior status, or jobs which had been reduced by downsizing.

Table 8.3 indicates the breakdown of fishermen's sources of reported income – wages only, wages plus benefits from unemployment insurance (UI), wages plus benefits from unemployment insurance and workers' compensation (WC), and benefits from The Atlantic Groundfish Strategy (TAGS). The distribution reflected the nature of the two fisheries. Deep-sea fishermen could work year-round for their companies, and about one-fifth of them did; however, 80 per cent collected UI during the summer months when their vessel underwent refit. In the coastal fishery only 10 per cent of our sample reported wages as their sole source of income. These men did not claim UI and had employment elsewhere in the off season. Two coastal fishermen decided to give up fishing and had taken the 'package' (TAGS). The remaining coastal fishermen claimed UI during the off season,' usually during the winter months of January through April. The higher levels of WC payments in the deep-sea fishery reflected the Workers' Compensation Board's practice of restricting WC only to fishermen who were employed in a contractual situation, and did not cover small boats of two or three persons.[1] It also reflected the higher levels of danger associated with the deep-sea fishery.

Other sources of income supplemented household finances, including Goods and Services Tax (GST) rebates, child support payments, family allowances, pensions, and rental income. Eleven coastal and seven deep-sea fishing households reported stocks and other investments as an addi-

Table 8.4 Distributions of Personal Earnings Reported by Wives of Deep-Sea and Coastal Fishermen (in percentages)

Wife's income Median	All coastal $10,000–$19,999	All deep-sea $10,000–$19,999
Less than $10,000	34.8	36.8
$10,000 – $19,999	33.3	39.7
$20,000 – $29,999	19.7	14.7
$30,000 – $39,999	4.5	2.9
$40,000 – $49,999	7.6	5.9

Missing values (0)

Table 8.5 Household Income Distributions Reported by Wives Broken Down by Household Types (in percentages)

Household's income Median	All households $40,000–$49,999	Coastal households $30,000–$39,999	Deep-sea households $60,000–$69,999
Less than $20,000	6.4	13.4	0
$20,000 – $29,999	11.4	15.6	7.5
$30,000 – $39,999	17.8	29.6	6.8
$40,000 – $49,999	17.1	14.1	19.9
$50,000 – $59,999	15.3	11.9	18.5
$60,000 – $69,999	11.0	9.6	12.3
$70,000 – $79,999	9.2	3.7	14.4
$80,000 or more	12.1	2.1	20.5

Missing values (0)

tional source of income. As we have seen, by far the greatest complement to a fisherman's income was wages from other household members, usually his wife's. Approximately 47 per cent of wives participated in the labour force: Table 8.4 summarizes the distribution of wives' reported income from full-time and part-time employment. The median reported income of employed wives for both coastal and deep-sea samples was $10,000 to $19,999. (Chapter 7 discussed women's employment in greater detail.) Table 8.5 indicates reported household income distribution for both samples. Including women's wages, the median reported income for coastal fishing-dependent households was $30,000 to $39,999, and for deep-sea fishing-dependent households $60,000 to $69,999. It must be noted that, in relative terms, wives of coastal fishermen contributed more to the household income than did wives of deep-sea fishermen. With the deepening of the fisheries crisis, erosion of fishermen's incomes (especially in the coastal fishery), and increased participation of wives in the workforce, women's wages had become proportionally more important in household finances. In some cases, wives' wages maintained the household's economy.

Household financial resources usually came from a wide range of sources, which reflected the needs of the household and the nature of the fishery. But, as we have seen, these resources remained unpredictable, as Opal's comments illustrate:

> In the wintertime, fishermen have it very, very hard. They could be tied up from October up until January or February, and you have to wait so many weeks for your unemployment [insurance]. You never know when the boat's gonna break down. Now, this trip they're home [for] two weeks, it's gonna be another week. They were sent to St Pierre, had a broker [made no money beyond expenses], never had no pay, they never made anything. So now he has to wait for the unemployment to kick in. And you go (sigh) three/four weeks, on the money you have saved – because you have to have so many months ahead if you have mortgage payments. You gotta have a little money in the bank, because when October comes, there you are waiting another four weeks, maybe longer, until your unemployment kicks in again. So, having my own money was nice (laughter). I didn't have to worry because I've seen trips where I've been left with twenty dollars. And that's to last me for twelve days!

Opal's situation indicates how one couple drew upon income from the fishery supplemented by unemployment insurance benefits and the wife's wages. For Opal, planning ahead and putting a little aside protected her household from the uncertainty of the fishery. In good times, her wages added insurance; in bad times, her wages became essential.

Why Plan?

In David McCrone's study of thirty-four households in Kirkcaldy, Scotland, he argued that 'while "strategies" exist in many households and provide broad prescriptions for action, they can best be understood not as neat blueprints but as ways of seeking to control and make sense of life's exigencies' (1994: 68). McCrone identified two household strategies: 'getting by,' the non-planning approach, and 'making out,' the long-term-planning approach. Planners sought to control and bring order to their lives. They developed strategies or plans that adapted to life's changing needs. Non-planners took things as they came. They tried to meet their needs head on and deal with the present. The future took care of itself. They coped one day at a time (69). These concepts are useful in examining financial management of fishing-dependent households.

Planning was not a universal characteristic of fishing-dependent households. Many households found budgeting impossible given the unpredictable nature of the fishery. Fishing-dependent households used both making out and getting by strategies. The following examples will illustrate the differences between these two approaches to household finances. Virginia and her husband, Victor, a coastal fisherman, had been married for ten years and had a child. Virginia, who worked for the household's fishing enterprise, explained her approach:

> I was always one for saving where we have no pensions. Like most people when they retire they get a pension from the company that they're working for, where we don't get that. I always felt that we had to make sure that we had our own when the time came for him to quit. Well, we basically know how much is going into the next paycheque because we know the price of fish and we know how many fish he has to sell. I've got it down to a system that I know exactly how much I'm going to need from one year to the next. Where I do the bookwork, I know how much money was spent on fuel, salt, and I knew how much money we needed to pay them bills during the year.

Virginia planned for the future. She recognized that sometimes she would not have any money to draw on, so she set aside money during the good times to draw on in the lean times. She drew up a long-term plan and worked to achieve it.

Karen, who worked in retail sales, and Ken, a deep-sea fisherman, had a more relaxed approach to household budgeting. As Karen, a mother of three children, explained:

> Whatever bills need to be paid, will be paid with whatever money's there. We're not picky about it (laughter). What's there's there, just pay for it. We don't worry about it. It's just money. It comes, it goes, it'll be there, it won't be there. Whatever we have is just 'things,' it's nothing. I mean, the house is wonderful. We like having the house, but it wouldn't be devastating if we couldn't afford the house. It's just a building.

Virginia tried to plan strategically; Karen got by. All households used coping mechanisms 'to get by,' but only some developed long-term strategies. Planning households made decisions at the short-term level, which would have long-term implications in the future. Households were more or less strategic, relying to a greater or lesser degree on coping mechanisms. Nor were these permanent approaches. For example, Karen and Ken planned strategically until Ken had an accident that almost crippled him. Before the accident Karen's motto was 'save, save, save for a rainy day,' but since the accident Karen has felt that 'Life's too short. You gotta enjoy it.'

But planning took two very different courses in these fishing-dependent households. In coastal fishing-dependent households there was an inherent struggle between the needs of the fishing enterprise and other household needs. This struggle played out between husbands and wives during the budgeting and planning process, and frequently resulted in financial conflicts. As Kristen stated, 'Sometimes when he's looking for a new boat (laughter), we usually have a few arguments about that, but he'll win. It's to do with his fishing, so he wins (laughter). I've resigned myself to the fact (laughter).' Any planning process in coastal households had to consider the needs of the fishing enterprise as well as those of individual household members and the household as a whole. In deep-sea fishing-dependent households the planning process only centred on the needs of individual members and of the household as a whole.

McCrone (1994) found that the distinction between planners and non-planners had 'a rough correspondence between access to resources and the propensity to "plan." Hence, those in more secure and better-paid occupations were more likely to be found among the "planners," as were men rather than women, and, within the latter group, women in full-time rather than part-time employment' (72). Following this logic, deep-sea captains' households would have a propensity to be planners while coastal helpers' households' would not. Although planning required having money, the nature of the fishery forced even the most impecunious household into some level of planning beyond a day-to-day coping strategy.

As we saw earlier, all fishing-dependent households had a common set of expenses: housing (rent, mortgage, taxes), vehicles (cars, vans, trucks), insurance payments (house, car), groceries, clothing, and monthly utilities (light, telephone, water, fuel). But coastal fishing-dependent households took on additional basic costs: mortgage and insurance on the boat, Goods and Services Tax (GST), and supplementary accident and health care insurance, since most coastal fishermen cannot get workers' compensation coverage. All fishers also paid directly to the federal government for fish licensing fees, unemployment insurance premiums, income tax, and Canadian pension premiums. In order to meet basic expenses, especially in coastal households, many households found it essential to have a budget and some level of short-term planning. Many wives of coastal fishermen used a squirrel analogy when talking about their household budgeting process. As Virginia once said, 'Like squirrels, you gather your nuts and then you save them for the winter. That's the way I always looked at it.' Once households met these basic needs, they had a number of competing demands on their finances. Long-term planners spoke of paying down mortgages and loans, saving for lean times, RRSPs, and their children's university education. Less long-term-oriented households saved for lean times or for special items such as a child's bicycle, new appliances, or a family vacation. Two households spoke of scrambling to make ends met, putting off creditors, and trying to hang on to their fishing enterprises.

The fisheries crisis forced households to reassess their financial strategies and to set priorities. Here's how Gilda described the effect of the fisheries crisis on her household's finances:

> You're used to making big money that you can contribute to an RRSP every year. You go to the grocery store, you buy whatever you want to buy. School time comes and you take the kids and you buy brand names. And now it's, get yourself a job because I can't afford your brand names. And you go to the grocery store, you buy all the sales, and you don't contribute to the RRSPs any more because you don't have any money to contribute, and you're worrying, do I have enough money to make my boat payment in September. You come down quite a bit and you change your lifestyle a lot. Not that you were a high roller or anything, but you just don't have thousand, a couple thousand dollars sitting in your bank account any more. You have to say, okay, next month's October, I've got the house insurance, and November's the car and truck insurance, I've gotta get some money saved somewhere, somehow, to pay those, because I'd die if I ever got behind on a bill. Like I could never stand to owe anybody any money, right? So it's quite different. And there's absolutely positively hardly any fish at all this year. It's not looking good.

Long-term planners spoke of having to draw down savings originally earmarked for retirement or their children's education to pay for daily expenditures, or choosing between supporting their children at university or maintaining their retirement funds. When asked if she had any savings left after putting their children through university, Nancy replied, 'No, we've blown it.' Short-term planners who were saving for 'lean days' used their savings to weather the crisis and many no longer had substantial savings. Some households approached the company, the fish buyer, local stores, or the family for an advance against future income, but short-term credit did not help their long-term budgetary problems.

In order to enhance household income, some women who had a job increased their hours; other women took employment, while still others were considering it. Everyone I spoke to talked about 'tightening belts' and 'making do.' Many spoke of limiting their credit risks, or refraining from using charge accounts or credit cards. Getting a loan was more and more difficult as banks became reluctant to extend loans to fishing families. But the crisis affected households differently. For some households 'cutting back' has meant reducing personal pleasures such as not going out for meals or movies as often. For others, cuts were more substantial and represented the abandonment of long-term goals, such as not putting away money for retirement or children's education, or deferring the purchase of a new car or truck or major appliance. Women spoke of doing more things for themselves around the home rather than using hired help. In some cases, men with substantial time ashore took on major repairs or maintenance of their homes.

For still others, with incomes so insufficient that investments and personal pleasures were beyond their means, the cuts were devastating. For these hard-pressed families, particularly in the coastal fishery, financial management meant setting up priorities among basic household expenditures and dramatically changing shopping habits. Women spoke of no longer buying new clothes and shopping only at second-hand stores, or sewing their own clothes, of making children's needs a priority while neglecting their own or their husband's needs, and of cutting back on grocery bills by eliminating 'all treats.' Their savings now went for short-term necessities such as children's Christmas presents or birthday gifts. In a few cases families found that they had to give up their homes, either selling their house to acquire equity for the fishing enterprise, or, if renting, moving to a more economical property. One young couple moved in with the wife's parents.

Who Has Access to These Resources?

Before examining household money management let's begin by looking at who had access to the financial resources of the household – bank

Table 8.6 Deep-Sea and Coastal Fishing-Dependent Households' Use of Different Types of Bank Accounts, Credit Cards, and Charge Accounts (in percentages)

Type of bank and other accounts	Coastal	Deep-sea
Chequing accounts		
Joint	78.0	81.3
Wife's	27.3	17.3
Husband's	20.7	11.3
Savings accounts		
Joint	70.7	78.7
Wife's	32.0	30.7
Husband's	22.7	23.3
Children's	78.7	58.7
Credit card*	72.8	76.7
In both names**	58.2	64.3
In wife's name only**	33.6	20.9
In husband's name only**	6.4	9.6
Have separate cards in own name**	1.8	5.2
Charge account*	68.2	62.4
In both names***	45.6	53.8
In wife's name only***	48.5	23.7
In husband's name only***	5.8	22.6

Missing values * (2); ** (35); *** (57)

accounts, charge accounts, and credit cards. (Table 8.6 summarizes these findings.) Households either pooled or segregated some or all of their financial resources. To ascertain access to these resources, and to see how households distributed their assets, involved looking at who was able to charge goods to a restricted account or a line of credit (e.g., credit card, charge account) and who could draw from a common account. More than three-quarters of all households in the study had joint chequing accounts; that is, 71 per cent of coastal households and 79 per cent of deep-sea households. About three-quarters of all households used credit cards, with higher participation rates in the deep-sea households. Approximately two-thirds of all households had charge accounts, with higher participation rates in the coastal households. In coastal households, 58 per cent of credit cards and 46 per cent of charge accounts could be used by both spouses. In deep-sea households about two-thirds of credit cards and charge accounts were in both names. Approximately 80 per cent of coastal households and 60 per cent of deep-sea households had set up bank accounts for their children, usually for post-secondary education.

When comparing these results, two patterns emerge. First, women in coastal fishing-dependent households had the higher frequency of exclusive access to accounts, credit cards, and charge accounts. Second, deep-sea fishing-dependent households had a greater frequency of joint, or husband-only, access to chequing accounts, credit cards, and charge accounts. These results suggest that wives of coastal fishermen may have had more control over household resources than their deep-sea counterparts.

However, access to pooled resources does not necessarily mean the ability either to manage or to control those resources. Nor does it mean real access. For example, although she and her husband had a joint chequing account, Martha did not have her name on the cheques. When asked why not, she replied: 'It's this man thing. He's got a chequing account – business – for the boat, and my name's on the account but he wouldn't allow me to have my name on the cheque. Yeah, I asked him, "Well, why?" It was, "Well, it just doesn't look right."' Although this reported interchange suggests a social reason for this situation, in reality it meant that it would be very difficult, if not impossible, for Martha to cash a cheque on this account. In turn, it meant that Martha's husband had control of both their finances and the coastal fishery enterprise, although here the wife was, in principle, a partner.

By looking at the frequency of joint and separate chequing accounts as indicators of pooled or separate resource management, I attempted to separate households into roughly four different types: joint access to pooled resources, wife-controlled access to pooled resources, husband-controlled access to pooled resources, and individual-controlled access to segregated resources. (Table 8.7 summarizes these findings.) These rough measures give us some indication of how households managed their money. Pooled resources indicate a potential for joint management, control, and access to resources. Segregated resources indicate individual management, control, and access to resources.

Based on whether or not a couple had a joint chequing account, I divided the samples in two groups: pooled and segregated. Most households had joint access to resources. Eighty per cent of all fishing households pooled their resources, while 20 per cent did not. In coastal households 78 per cent of households had joint access to financial resources, while in deep-sea households 83 per cent had such access. I

Table 8.7 Use of Segregated versus Pooled Resources by Deep-Sea and Coastal Fishing-Dependent Households (in percentages)

Type of account	Coastal households		Deep-sea households		All households	
	N	%	N	%	N	%
Pooled resources						
Joint only	85	56.7	101	68.2	186	62.4
Joint plus wife's account	20	13.3	13	8.8	33	11.1
Joint plus husband's account	12	8.0	8	5.4	20	6.7
Segregated resources						
Wife's account only	14	9.3	9	6.1	23	7.7
Husband's account only	12	8.0	13	8.8	25	8.4
Wife and husband with two separate accounts	7	4.7	4	2.7	11	3.7
Number of cases	150*		148**		298	

Missing values * (1); ** (2)

then broke down those households with pooled assets into three categories: joint chequing account only, joint chequing account plus wife's own chequing account, and joint chequing account plus husband's own chequing account. (There were no households with both a joint chequing account and separate accounts for each spouse.) Sixty-eight per cent of deep-sea households and 57 per cent of coastal households had only joint chequing accounts. These households had the greatest potential for being jointly managed and controlled. In coastal households 13 per cent had a joint chequing account, with the wife maintaining a separate chequing account; and 8 per cent had a joint chequing account with the husband maintaining a separate chequing account. In deep-sea households 9 per cent had a joint chequing account, with the wife maintaining a separate chequing account, and 5 per cent had a joint chequing account with the husband maintaining a separate chequing account.

Now let's look at those households that did not have a joint savings account. I separated these households into three categories – only the wife having a chequing account, only the husband having a chequing account, and both husband and wife having a separate account. In approximately 8 per cent of both coastal and deep-sea households, only husbands had a chequing account. In 9 per cent of coastal households, only wives had a chequing account, and in 5 per cent of these households, each spouse retained separate bank accounts. In 6 per cent of deep-sea households, only wives had a chequing account, and in 3 per cent of these households, each spouse had a separate bank account.

How Are Decisions Made?

Access to resources indicates, at one level who controls resources, but only indirectly. In the survey I asked wives: 'Who is in charge of the household accounts?' and offered them three responses: I am solely in charge; we are both in charge; and my husband is solely in charge. Table 8.8 summarizes whom wives reported as managing the household accounts, and indicates how these responses break down by households that pooled or segregated their financial resources.

Whether the household pooled or segregated resources, for both samples women most frequently managed the money (62%). In coastal fishing-dependent households 70 per cent of pooled resources households and 66 per cent of segregated households, women were the principal money managers compared to 55 per cent and 54 per cent, respectively, for deep-sea dependent households. In deep-sea house-

Table 8.8 Management of Household Accounts by Wife and/or Husband in Deep-Sea and Coastal Fishing-Dependent Households (in percentages)

Who Manages	Coastal			Deep-sea			All households		
	N*	% of group	% of total	N**	% of group	% of total	N	% of group	% of total
Pooled resources									
Both	25	21.4	16.8	40	32.8	27.0	65	27.2	21.9
Wife	84	71.8	56.4	67	54.9	45.3	151	63.2	50.8
Husband	8	6.9	5.4	15	12.3	10.1	23	9.6	7.7
Number of cases	117		78.5	122		82.4	239		80.4
Segregated resources									
Both	7	21.9	4.7	8	30.8	5.4	15	25.9	5.1
Wife	21	65.6	14.1	14	53.8	9.5	35	60.3	11.8
Husband	4	12.5	2.7	4	15.4	2.8	8	13.8	2.7
Number of cases	32		21.5	26		17.7	58		19.6

Missing values * (2); ** (1)

holds a greater number of men managed the money than in coastal households. Once again, these findings indicate that women in coastal fishing-dependent households had greater access to the household finances than did their counterparts in the deep-sea fishery.

Yet being in charge of household accounts did not necessarily mean making decisions about how the money would be spent. It may merely have meant being the bookkeeper and the cheque writer. For example, when I asked Unice who was in charge of the household accounts, she answered, herself, yet when pushed to explain what this meant, she responded:

> I pay all the bills, but we do our banking together, we make our decisions together. Oh yes, if we need new things or anything we do that together. But, like I said, I pay the everyday bills, light bill and the phone and all this stuff, fuel ... If there's any decisions to be made, investments or anything, we do [it] together – definitely not by myself. I wouldn't have the nerve for that. I mean, it's just a thing that I think a married couple does. I mean, I have the free hand with all the money that I want to spend and what I want to do with it, but just certain things, major purchases, that we decide on ourselves. If we need anything done to the house or anything, we decide between us.

Other women talked about doing all the household management but ultimately not having control over the decision-making process. Zoey's comments about her first husband best summed up the process: 'If he needed something no matter what it was he would buy it, but if I wanted a new TV, I would say, "Well, we need a new TV. We spend a lot of time watching TV," and he'd just say, "Ah, it's good enough." And I wouldn't go out and buy it unless he agreed, which most of the time he didn't. But if it was a truck for him, a new truck, or something like that, that was no problem, he'd have the new truck.'

Another way to look at financial control was to see who made the decisions about purchases. Table 8.9 summarizes how decisions were made concerning purchases. Coastal and deep-sea households differed in daily provisioning and in making of major purchases. In coastal households wives had more control over the decision-making process regarding daily purchases. Almost two-thirds of all wives had sole control over daily purchases, and 35 per cent of all coastal wives said they had no limits on this spending. In comparison only a third of deep-sea wives had sole control over daily purchases and less than one per cent said they

Table 8.9 Spouses' Decisions on Daily and Major Purchases (in percentages)

Who Decides	Coastal daily purchases	Deep-sea daily purchases	Coastal major purchases	Deep-sea major purchases
Wife only	64.9	32.2	3.3	0
Wife mostly	8.6	23.1	17.2	11.3
Both spouses equally	20.5	35.5	54.3	67.3
Husband mostly	4.6	0.8	21.9	14.7
Husband only	1.3	4.1	2.6	5.3
Varies when husband is at home or away	0	4.1	0.7	1.3
Number of cases	151	150	151	150

Missing values (0)

had no limits on such spending. Whether major or minor purchases, more deep-sea households made more joint decisions than coastal households. But when making major purchases more coastal husbands made financial decisions alone, or mostly alone, than deep-sea husbands. Table 8.10 uses the question 'Who makes major purchases?' as an indicator of who controlled the decision-making process, and breaks down these results by pooled and segregated resources. As expected, in segregated households individual control of resources appeared higher than in pooled resource households. Deep-sea households seemed to have more joint control of resources.

Owner/operators in the coastal fishery inherently made more decisions on their own because the fishing enterprise drew upon the household's finances, and fishermen primarily made decisions about the fishing enterprise. Because the fishing enterprise sustained coastal fishing households, the needs of the boat had priority over all other household needs. Although consulted in these matters, wives had no real veto. As Unice said, 'Yes, I'm always consulted on gear and the boat and things like that, and I'm sure he can make those decisions perfectly well without me, but he does always check to see what I think about it.' Like Unice, most coastal fishermen's wives recognized that the fishing enterprise had priority, and unless the financial drain of the boat became so crippling, as it did in Laurie and Len's case (discussed in chapter 3), the wife continued to struggle to pay for the household necessities out of whatever income remained. Sometimes households could not get by. As Janet mentioned, 'When he was fish dragging we practically went bank-

Table 8.10 Wife's and/or Husband's Control of Household Accounts (in percentages)

Who Controls	Coastal			Deep-sea			All households		
	N*	% of group	% of total	N**	% of group	% of total	N	% of group	% of total
Pooled resources									
Both	66	56.4	44.0	87	71.3	58.8	153	64.0	51.3
Wife	22	18.8	14.6	11	9.0	7.4	33	13.8	11.1
Husband	29	24.8	19.3	24	19.4	16.2	53	22.2	17.8
Number of cases	117		77.9	122		82.4	239		80.2
Segregated resources									
Both	16	48.5	10.6	14	53.8	9.5	30	50.8	10.1
Wife	8	24.2	5.3	6	23.1	4.1	14	23.7	4.7
Husband	9	27.3	6.0	6	23.1	4.1	15	25.4	5.0
Number of cases	32		21.9	26		17.6	59		19.8

Missing values * (2); ** (1)

rupt. We almost lost the house.' Two fishermen took the TAGS package after they went bankrupt and lost their homes.

But once again the rhetoric concerning the decision-making process did not reflect the reality. In coastal fishing-dependent households I attempted to separate the decision-making process related to the fishing enterprise from that of making major purchases for the home. To do this, let's return to an example from the previous chapter. Trudy said that she and Tim jointly made decisions concerning major purchases for the household, but that she deferred to him on major boat purchases. As Trudy said:

> We were just doing that last night. I'm going to write down on a piece of paper what we want to get with this fall's good fishing, and the minimum that we want to save, and then what we want to do. We'll have that all down on paper, and we'll check it periodically to see how we're doing, against our projected plans. He wants to get a new chesterfield set for the living room. I'm completely happy with what I have (laughter), but he says it doesn't hug him (laughter) when he sits down. I would like to keep it and get it re-covered – I can do part of the work myself – but I just said, okay then, you want to get a new chesterfield set, that will be one of the things that we will buy this fall. And he wants a few pieces of equipment for the tractor; they'll come to about three thousand dollars with maintenance, so we'll see that we get that. And I think that was all. We want a CD system for the living room. We have one in the car, but (laughter) I don't do enough travelling in the car to listen to all the CDs, so I think that will be possibly all that we'll spend money on for major things this fall.

Trudy did the financial management of the household and looked after the household's joint chequing and savings accounts. Her first priorities after paying the basic household expenses included long-term savings for their pensions and their children's education, and buying equipment for her husband's tractor. Although she and Tim agreed on these priorities, they differed with regard to their next purchases. She wanted a CD player for the living room; he wanted a new chesterfield set. For the same amount of money as the new chesterfield set she could have the set they had re-covered and tightened and still purchase the CD player, but she deferred to him. When pressed later in the interview about this, she replied, 'Well it is his money.'

Differences in findings concerning access, management, and control of a household's finances might merely reflect the needs of a fishing enterprise. At one level these results seemed contrary to expectations. If deep-sea fishermen remained at sea most of the time, then one might expect that their wives would have more control over the household accounts than coastal fishermen's wives whose husbands came ashore more frequently. But this expectation would be wrong, for several reasons. First, generally speaking deep-sea fishermen made good money; even if their wives had a job, they usually made substantially more money than their wives. This inequality in contribution to the household finances compromised women's entitlement to household resources. As we have seen, many wives saw their husband's wages as 'his money.' Second, husbands went to sea for relatively long periods of time, and, as some women said, their husbands had an inherent lack of trust in them and their management skills, especially when they were first married. As a result, the wives felt they needed to account for their expenditures while their husbands were away. Third, wives spoke of the need to integrate their husbands into the running of the household, to make them aware of what their money bought, and thus demonstrate to them what good providers they were and what good managers their wives were. By integrating husbands into this aspect of the household, wives hoped to integrate them more into their households' other concerns and interests.

On the other hand, in coastal fishing-dependent households work – paid and unpaid – was more integrated. Husbands frequently relied on wives to make daily financial decisions concerning the fishing enterprise, such as the repair of a boat part or the price of fish. Over half of these women looked after the fishing enterprises' books and accounts. Wives who acted as shore skippers could not necessarily confer with their husbands about every purchase. Husbands had to rely on their wives' judgment to make the best deal on their own. Husbands then developed a high level of trust in and respect for their wives' financial and accounting abilities. A wife's work in the fishing enterprise made her a partner, albeit a junior one in many cases, with a stake in the enterprise's financial well-being. Moreover, the difference in income generated from women's employment and that of their husband's income did not differ substantially, as it did in deep-sea households. These women felt more entitled to managing and controlling the household's finances. In fact, in some cases the wife's wages kept the fishing enterprise afloat, as well as supported the household. Moreover, women's unpaid labour in the fishery made yet another economic contribution to the household enterprises.

178 Set Adrift

How Do These Households Work?

In the sections above I separated out various components – access, management, and control – of household financial resources strategies. But how did these various processes relate to each other in the households' functions? Using interview material, I have identified types of financial management strategies – housekeeping or allowance-only systems, and pooled or segregated system – used by fishing-dependent households.[2]

In single male wage earner households for both samples, by far the most common method of management of finances involved a wife's control/management of household finances, with her husband having an allowance for his personal needs. In this situation, the wife did not have paid employment, but she managed and controlled the finances in ways she equated with power and self-esteem. As Laurie said, 'It gives me my sense of power to know that it might not be my money but I'm still in control of it.'

In the deep-sea fishing-dependent households, this management system allowed the wife access to the household finances while her husband fished. Some women, such as Cathy, argued that the organization of their work and the work schedule of their husbands' job meant that they looked after the bills. According to Cathy, she just fell into the habit of doing the bills. Bills came due when her husband was at sea, so she paid them; bills came due after he returned home but he was too busy to do it, so she paid them. Other women argued that their husbands were 'no good with finances.' As Amy said, 'If he was controlling the finances, well, we'd probably have been on the sidewalk a long time ago. Because money burns a hole in his wallet, and that's why I keep control of the finances.' But whatever the reason, most of the women who used this system claimed that their husbands had never written a cheque nor received one. Take Laurie's and Len's household as an example:

> Len always makes the comment he has never seen a cheque. I shouldn't say never; I don't know how many settlements he'd have a year, but if he'd see six settlements a year, that would be it, six cheques a year, that's his maximum. Usually they come mailed home to me. There's a settlement sheet, where he's the captain. He could just look that over. But as to the dollars and cents that he was exactly paid, very seldom does he know what it is unless he calls home. And he doesn't call, he just says, okay, Laurie knows what's best and what to do with it and where it belongs. Usually no questions asked. He knows that I'm not going to go out and refurnish the house

or do anything crazy like that. I'd love to (laughter), but anyway I wouldn't. But no, I'm in charge of the finances. He has a credit card and he has a limit (laughter).

Laurie's comments illustrate her pride in her ability to manage the finances, and the recognition that her husband saw her as a capable manager. But not all the women were appreciated in this way. For example, Vera felt that her husband did not appreciate the difficult job she had trying to balance their finances. Although Vincent made good money, Vera needed to keep him on an allowance. As she explained:

If things were kept up with, like we weren't behind in anything [things were okay], but there were times I'd have to sit there and count change. But he just didn't seem to understand that he couldn't take five or six hundred dollars every paycheque. I don't know what he'd do with it (laughter). But, yeah, that was kinda hard there. And we used to fight about that a lot, because it was his paycheque and he just couldn't understand why he couldn't take what he wanted out of it. And when I'd say, 'Fine, you do the finances,' [he'd say] 'Well no' (laughter).

Disputes surrounding budgeting represented the most serious disadvantage to the wife's management or control of finances. Many of the women whose households used this system spoke of their need to explain to their husbands how the accounts functioned. Some felt it was unfair for them to have to shoulder the responsibility, yet they were loath to give up managing the finances. In coastal fishing-dependent households where the wife managed the household finances, she might have managed the fishing enterprise finances as well. Many women who willingly took on the responsibility of the household finances balked at the additional burden of doing the fishing enterprise's books.

Another common type of management we have seen involved the fisherman's controlling and managing the household. In the most extreme variant of this system, the husband took on all the responsibility for managing household resources and expenses except for his wife's personal spending money or an allowance. In the other and more common variant of the system, the husband took on all the responsibility for managing household resources and expenses, but gave his wife not only a personal allowance but also a housekeeping allowance to pay for the common expenses of running the household day-to-day. In the latter situation, the husband and wife negotiated a sum needed to maintain the household, and the wife then took over responsibility and control of this money to meet their needs.

To illustrate the variants of the system and how it functioned, consider the differences in the respective management schemes of Donna and Dean (housekeeping system) and Paula and Peter (allowance only system). Paula and her husband, a deep-sea fisherman, had set up the following scheme, which allowed Peter to have control of the entire household finances while enabling Paula to cope on a day-to-day basis.

> He goes and gets his cheque. He has his own separate chequing account and savings account that the mortgage payment comes out of, and then he gives me the cheques to pay the bills. So I very rarely ever see his cheque, just because I'd spend it. When he goes out to sea, every bill is paid. There's groceries in the house for while he's gone. There's diapers, there's formula, everything I need while he's out is here ... I don't need anything. And then he gives me a couple hundred dollars for the two weeks that he's out, just for whatever I want. If I want to go out with my friends or if I want to take the kids out to dinner, or whatever, I always have money on hand if something comes up. And I usually have a blank cheque to his chequing account, in case of emergency.

In this example, Paula, who stayed home and looked after their three boys, depended totally on her husband for all financial resources. Peter alone had access to the bank accounts and the credit card. He even did the grocery shopping with Paula, paying the bill at the checkout counter. Paula had no income, only the personal spending money Peter gave her. Paula also spoke of receiving additional sums of money from her husband designated for specific purchases such as clothing and Christmas presents for herself and the children. In most of the households identified as using this allowance system, the wives did not have paid employment, had no external sources of income, and depended solely on their husbands for all of their monetary needs.

Donna and her husband, Dean, a deep-sea fisherman, ran their household finances this way: 'Well, he [has a bank account], and then we have a joint account. The way this company works where they're gone for so long, halfway through the trip the cheque is sent out – not their full cheque, but [an advance]. I take a percentage of that for the month, and then put the rest in his account. So I kind of have this budget to live on. I guess it was kind of a mutual thing. We just sat down and figured out the major bills, the definite bills that we have to pay, and then how much I need, like, for gas and groceries, and then allowing me a little extra for [things] like the movies (laughter).' Donna relied on her husband for financial support, but unlike Paula she had some financial responsibility for the household.

In deep-sea fishing-dependent households, there were inherent structural difficulties associated with husbands' managing and controlling all the household finances. For example, what happened if an emergency arose when the husband was away? Deep-sea fishing-dependent households used the housekeeping variant of this management system because it allowed husbands control over most of the household finances while giving wives the flexibility to manage the daily expenses generated by the household while the husband was away. In the survey, these households appeared as pooled resource households. Although many of these women had joint chequing and/or savings accounts with their husbands, they did not have access to nor did they manage the household's finances beyond daily provisioning.

In the coastal fishery, husbands who managed all household resources usually managed the fishing enterprise, too, which allowed them full access to and use of all household resources for the fishing enterprise. In coastal households the use of the housekeeping system assured the family of having some money assigned to daily provisioning and household expenses. Under the system of male control and management, the needs of the household became secondary to the needs of the fishing enterprise. By setting aside funds clearly marked and assigned to household expenses, as being under the control and management of the wife, the fishing enterprise could not subsume all of the households' financial resources. It also allowed husbands to hive off their fishing business and run it separately from the household while maintaining an interdependency of household and fishing enterprise spheres.

The households of the newly married or those with young or dependent children used this system most. During Quinsee's first marriage to a deep-sea fisherman, the couple managed their household under the housekeeping system. 'When he was home, he would pay all the bills and put in the bank whatever he wanted to put in the bank. And whatever was left for spending those two weeks or whatever, that's pretty much what we went by.' Later I asked Quinsee, why she thought they did it that way. She replied, 'When you're first married and you don't know the other person really that much, and you're making big money like he was, well it was a lot of money in them days – forty, fifty thousand dollars a year scalloping.' Husbands' lack of trust in their wives' ability to manage the household finances underlay both Paula's and Quincee's experiences. Paula had a history of overspending when she had her own credit card. Quinsee explicitly indicated her first husband's lack of trust in her financial ability, but it went even further: when he was at sea, he tried to curtail her activities by insisting that she account for every penny she spent, including her personal spending money. Thus her ex-husband used access to money and its expenditure to modify her behaviour. This coercion contributed to the breakup of their marriage.

In other cases, wives reported that they felt they had no right to manage the household funds: 'Well, it's not my money as far as I'm concerned. I mean, he's working for it, I'm not.' Most of these women had little or nothing to do with managing the household finances; others would help by getting the bills together and physically dropping off the cheques. Wives did have some say in how the household income was spent: for example, Paula would ask for money for the children's clothing or a new washer and dryer, but the final decision concerning where the money would be spent remained in the hands of her husband. Since women felt they had no access to money except through their husbands, they had to comply.

Over 45 per cent of the women interviewed participated in the labour force, however, and did contribute to the financial resources of the household. As we have seen, there appeared to be two methods for incorporating such wages into the household. One method involved the wife's controlling and having responsibility for as many of the expenses related to daily provisioning and household expenses as she wanted to take on. The other involved the household's pooling its resources and then drawing from these enhanced resources for all expenses. The former freed up money from the domestic budget, usually resulting in additional funds under the husband's control. It most often occurred when women had only part-time or seasonal jobs, many of them paying minimum wage or just a little bit more. The latter strategy allowed more money to be available to both husbands and wives, and occurred most often when wives had full-time and relatively well-paying jobs.

When incomes between husbands and wives were unequal and finances were precarious, the most common method for managing household finances appeared to be for the wife to use her money for both her personal expenditures and some if not all of the household expenses. In coastal fishing-dependent households, management responsibility divided between the household and the fishing enterprise, with the wife's taking as much responsibility as possible for the household and the husband's taking full responsibility for the fishing enterprise. For example, Trudy and her husband, Tim, used a housekeeping system until Trudy started receiving income from the rental of a house she had inherited. Trudy used this money for both her personal expenses and some of the household expenses. Since she brought in additional financial resources to the household, she did not require a personal allowance from Tim. This additional source of income allowed him additional flexibility and gave him more resources for the fishing enterprise.

Quinsee and her second husband, Quentin, also used this type of management system in their household. Quinsee had a part-time job, and they initially pooled their resources. However, because there were too many conflicts over finances, they moved to a system in which each person became responsible for a particular set of bills, thus effectively keeping their finances separate. Quinsee used her money to pay for the daily housekeeping expenses and monthly household bills. Quentin paid for the expenses associated with the fishing enterprise and large household expenditures such as the utilities. By using this management method Quinsee felt much more independent and had more self-confidence than in her previous marriage. She spoke of being able to make decisions on her own for the first time, and feeling pride in this accomplishment:

> If we need to buy something – like right now we're talking about buying a new washing machine – we usually just buy it. And if he's gone when it comes to the point that it does go, I'll probably just go pick up one and get it sent home. I can't go very long without a washing machine, but, well, we usually talk to each other and decide on anything like that. We never have a problem. He's the type of person that he'll buy whatever he thinks he needs and if I think it's too much or if I don't think we should have it, I'll just say no, we don't have the money right now, and he'll just go with the flow. Never have too much problem with him. I got him well trained, I think.

Throughout the interview it became apparent that Quentin and Quinsee did attempt to make joint decisions about their household expenditures, with Quentin deferring to her. It was less clear how much say she had in the fishing enterprise. However, having her own money and contributing to the household finances gave her a sense of power and independence.

Finances usually were pooled when wives had full-time employment and their wages were relatively equal to or more than that of their husbands. Such wives frequently managed all household accounts, and/or made most of the decisions pertaining to the household, only consulting their husbands concerning major purchases. In coastal households husbands might manage all fishing enterprise accounts, but consult their wives about the business, especially if 'her' money would be used for these expenses.

Women in deep-sea fishing-dependent households with pooled resources from two 'good' incomes were most likely to have equal control over the household finances and to be the household managers. For example, as Gail explains, she and Gordon managed their household in the following way:

> In our household I take care of all our financial kind of things completely, even some of the major purchases, or decisions that have to be made. I don't wait for Gordon to come home. I think, well, should I or shouldn't I do this? Basically, if something has to be done [I do it]. Like last week, I needed four new tires for the car. Well, I wasn't going to wait for him to come in to make the decision (laughter). You decide, go purchase them, have them done, or whatever. He finds out basically after the fact, or whatever. It's basically luck, because he trusts my judgment that when I make a decision, it's going to be a good decision.

Once again the prevailing attitude seemed to be trust in the wife's judgment.

But some households did not fit easily into any of these systems. For example, Oriel and Obadiah pooled all their resources, but Oriel had full responsibility for the household and Obadiah managed the fishing enterprise. As Oriel recounted:

> I manage all the household finances, everything that's related to the house and our spending money. Anything that's related to his business, he handles. I have no idea of what the business makes or doesn't make, or what's in the business accounts – nothing. I have no clue. I set that up from the time we first got into it because I didn't want the responsibility. Just managing the household money was enough, because when you fish you don't get regular paycheques like most people do, so you would get a lump sum, but it might have to do you for three to five months. Plus, hard enough trying to manage and make sure you got all the payments down pat and whatever, so I could not handle doing his business.

In this household, Oriel not only had the money from her job but also access to all of her husband's income, except for the money needed to run the business. Obadiah got the boat's share – usually 25 or 35 per cent of the profits – for the fishing enterprise, and Oriel got Obadiah's share: 25 per cent of the profits. But she had no way of knowing when she would receive the money from Obadiah or how much money she would have: 'And I don't question it. I don't ask. I just say, okay, fine, that's what I get. If I get it this week or if I get it once every three months or whatever, I'm fine, every bit helps.'

Oriel saw their system as a win/win situation, where she maintained her independence and her husband had less stress. She continued to explain:

> Well it makes it easier on him, like he doesn't have all the pressures of bringing in all the income. So, if he has a bad month or a bad season, at least we know that we have my paycheque as a steady paycheque. And that's how we've always geared our life to it. We don't live beyond our means, really, and we don't really have to worry about too many bills because we live within our paycheques. We lived totally on my salary for five years until we built his business. So I know what it's like to scrape by and survive, because it's my independence and also because we've grown accustomed to having two salaries. Even in the good times and bad times you get accustomed to a certain lifestyle and you like to always maintain that. Sometimes you can't, but at least you've gotta be thankful that we have, like I say again, one good steady income coming in the door.

Oriel's income freed up Obadiah's fishing enterprise from household finances, and her management of the household expenses also eased his worry about managing two sets of accounts.

CHAPTER NINE

Conclusions

I wouldn't change him one bit. We look like we come from two different worlds, and a lot of people think that, but he has a lot of special, tender things about him that makes him special to me. And I'm very lucky to be married to a very easy-going, laid-back individual when he's on his own time. But when he's in the fishing industry and making his living, he's a hard worker trying to survive in an industry that I have my doubts about (laughter), but he doesn't. They're all that way. They all believe that it's going to turn around and the next trip they're going to get bigger and better fish. And you can't discourage them no matter how hard you try, so I don't even attempt it. If that's what makes you happy, then I'm happy. And you accept that.

But, personally, the fishing industry robs people of the everyday lives, because they have to work away so much that you don't have a normal, everyday family life. You know people take for granted getting up every morning and travelling to work together. I've never experienced that. I've never experienced talking over everybody's day, at the end of the day travelling home together. Cooking supper and eating it all at the same time, that doesn't happen in our house on an average week. We might maybe get two nights a week that that happens. Those are things that I sometimes wish we could've had, but I've accepted what we do have, and it's got its good points, too. You know, that's an everyday humdrum life, but to me we've never experienced it, so we don't know.

I think it will always be a good way for the man that loves it to make a living. He'll always love it. You just have to accept it. If you want to stay married to them, that's what you have to accept. If you don't want to be married to them, then you just walk away from it and let them go. I've put too many years into it to do that (laughter). Anyway, that's it. That's my final comment. (Oriel)

I have begun this chapter with Oriel's concluding remarks because they capture both the passion and pain of fishing-dependent households. They summarize the reasons why men fish for a living and why their women stay and support them in their endeavours. Whether coastal or deep-sea, fishing-dependent households continue to strive to make a living from the oceans, in some cases under dire circumstances.

In the preceding chapters I noted how the structure, organization of work, and demands of fishing shaped the day-to-day lives of members of fishing dependent-households, particularly those of women. Coastal households confronted challenges specifically related to their business enterprises. For these couples, variations in household needs and responsibilities competed with the changing demands and obligations of the households' fishing enterprises. The degree of integration of each couple's work spheres depended on the husband's need for his wife's labour, his willingness to have his wife work with him, and the willingness of his wife to participate in their household's fishing enterprises. Demand for the wife's labour depended on the type of fishing the household engaged in, and on the husband's job status – as crew, captain, and/or owner of the vessel. For many coastal fishing-dependent households, increases in the wife's responsibilities related to various stages in the life cycle and to other household demands that paralleled the husband's increasing responsibilities and the demands of his career as he moved from crew to operator of his own boat. These competing demands between the wife's domain and the husband's domain framed the daily negotiations in each household.

Deep-sea fishing-dependent households also confronted specific challenges related to their type of fishing. These fishermen's work took place in an industrial setting, determined by the demands of the fishing companies, the nature of the fishing endeavour itself, and the workers' career cycles. The constraints of the deep-sea fishermen's work schedules defined the organization of each household. The degree of integration of each couple's domain depended on the way the company organized the husband's work schedule, the wife's organization of her work both inside and outside the home, and on the willingness of both spouses to participate in household activities when the husband returned home. Unlike coastal fishing-dependent households, where men's and women's work spheres overlapped, the sexual division of labour in deep-sea households set up two separate domains, which intersected only briefly when the men returned home. In these homes, women's work included the daily routine of running the household, with specific needs and responsibilities varying depending on the family life cycle: newly married, raising children, facing retirement. Almost half of the women in this study were employed outside the home, and the needs and demands of their jobs also had to be incorporated into their daily work plans.

Whether the men fished the deep-sea or the coastal waters, the fishing industry could not survive without the labour of these women. The women married to coastal fishermen contributed to the household fishing enterprise through their direct participation in fishing activities and through their business and financial management, their support services, and their domestic labour. The women married to deep-sea fishermen contributed through their domestic labour, maintaining the households while their husbands were at sea, and caring and giving support when they come ashore. By revitalizing their men during shore leave, these women prepared their husbands both mentally and physically to return to the sea. Thus women's unpaid labour in the home and in the fishing enterprise filled in or replaced other forms of labour. These livelihood strategies support Elson's (1992: 26) argument that women's unpaid labour is crucial to each household's ability to absorb the adverse effects of economic restructuring.

Moreover, with the deepening of the fisheries crisis, women's employment outside the home and their concomitant contribution to their households' income had become crucial in maintaining the fishery itself. This was especially true both in coastal and deep-sea households where fishermen were laid off, because women's wages then became the mainstay of these households' finances. Once again, these livelihood strategies relied on women's labour, in this case women's wages, to alleviate the adverse affects of the restructuring of the fishing industry.

But this study goes beyond the comparison of adaptations of fishing-dependent households. It reveals as well that domestic labour, primarily women's work, sustained both types of households, and that these day-to-day activities reflected the underlying power dynamics within each household. Helping out and other forms of reciprocity and exchange allowed women to meet their responsibilities and obligations. They depended on others – husbands and other kin, particularly female relatives – to help meet the daily challenges, but their dependence on others went beyond their families and included their social networks as well. Each individual woman's social network changed dramatically throughout her lifetime, reflecting the various needs that arose. As household responsibilities and obligations changed through time, and/or as these women took on paid em-ployment, their needs for various types of support also changed. In part, changes in women's social activities and friendship networks reflected the changes in their households. But social networks not only formed a mechanism for exchanging goods and services, they also became foci for social activities and agents for and against social control. In this study these social constraints were reflected in the gender and occupational segregation of friendship networks common in both coastal and deep-sea fishing communities. Some women did, however, break away from these constraints and develop friendship networks beyond the fishery.

One aspect of social relations – the drinking of alcoholic beverages – had a specific impact on these households. The organization and working conditions of the Nova Scotia fisheries, particularly in the deep-sea fishery, combined with the stress inherent in working at sea, provided a structural framework and an environment for both high-risk and low-risk binge drinking behaviour. The additional stress associated with the fishery's decline exacerbated the situation. For many of the wives affected by this problem, managing their husbands' drinking became part of their caregiving role. By ensuring that a husband could control his drinking, a wife felt assured that he could fulfil his obligations and responsibilities to his employer or to his own fishing enterprise, as well as to his household. Some effective strategies included wives' helping to integrate their husbands more fully into family and community life, and helping their husbands develop interests outside of fishing (and its related drinking).

Women who joined the labour force faced a myriad of challenges. In an effort to balance their household and employment responsibilities, many employed women ended up doing a 'double or triple day' of labour. And the responsibilities of home and workplace did not remain constant: family responsibilities changed through time and so did employment demands. These challenges are common to all employed wives, of course, but these fishermen's wives had to cope as well with the additional demands of their husband's work and/or their households' fishing enterprises.

The financial resources available to fishing-dependent households, and the ways household members generated and used these resources, were important in understanding the power relations between husbands and wives. In other words, assuming that money is a source of power, the relative economic position of husbands and wives mirrored their household relationships. Fundamentally, the differences between households that adopted 'a just getting by' approach and those that attempted long-term financial planning reflected the level to which the fisheries crisis impinged on the household's financial well-being. The fisheries crisis made new and in many cases harsher demands on the latter than on the former. The crisis also had in some cases challenged and in other cases undermined the rigid sexual division of labour so characteristic of the fishery historically. In the deep-sea fishery, which conventionally reinforced the sexual division of labour, double crewing of vessels had in some cases mediated this separation of men's and women's work. In the case of the coastal fishery, which by its nature sets up more opportunities for shared work between spouses, the fisheries crisis led either to more involvement or less involvement of women, depending on the strategy employed by the fishing enterprise. In some cases, women had more opportunities to participate in the fishery enterprise, particularly in those households, which depended heavily on the inshore lobster fishery. But in other cases, where women now worked in paid employment or where their husbands fished further from home without regularly returning at night, women participated less in the household fishing enterprise and more in the day-to-day running of the household.

What of the Future?

The crisis in the North Atlantic fishery of the 1990s led directly to a restructuring of the fishing industry and its fishing-dependent households and their communities, in turn sending reverberations throughout Atlantic Canada. The fishery could no longer be the economic mainstay of many coastal communities. But what would take its place?

The controversial report *Economic Adjustment in Selected Coastal Communities* (CIRRD 1995) proposed that only a few centres outside of the metropolitan areas in Atlantic Canada would survive the economic crisis brought on by the collapse of the fisheries. This report looked at two communities in southwest Nova Scotia – Shelburne and Lockeport. This study did not review the Lunenburg/Bridgewater area, but many of the report's comments made about the decline of the economic base of the fishery in the Shelburne area, a community about 100 kilometres further south along the coast, apply equally to the Lunenburg/Bridgewater area. Although the latter area has a few manufacturing plants, including Michelin Tire, the decline of the fishing companies based in Lunenburg, such as National Sea, undermined the economic base of the town and surrounding area. The municipal government and community leaders targeted Lunenburg's old town and the adjacent shoreline for expansion of the tourist industry, and possible development by the film and television industries.

Lunenburg, designated a National Historic Site in 1992 and a UNESCO (United Nations Educational, Scientific and Cultural Organization) World Heritage Site in 1995, has a rapidly growing tourist trade. In fact, tourism was well on the way to replacing the fishery as Lunenburg's primary economic base when the crisis struck. Throughout the mid and late 1990s and into the twenty-first century the metamorphosis from fishing centre to tourist designation accelerated. The process of becoming a tourist designation involved the creation of a product, which could be bought and sold (Cohen 1988; MacCannell 1989). In the case of Lunenburg, this commodification focused on its historical ties to the sea and particularly to the fishery. Through the process of commodification, the historical and culture entity known as Lunenburg was transformed into a distorted reflection of itself. It claimed authenticity and yet it was not (Cohen 1988; Urry 1990).

The tourist transformation of Lunenburg focused on the 'old town' and included not only the gentrification and commercialization of historical buildings and other aspects of the physical landscape, but also modifications to the social and cultural landscape. The blacksmith who once made anchors and other hardware for fishing vessels began creating decorative iron works (e.g., light-switch plates, andirons for fireplaces) to grace gentrified homes. With the revitalization of boat building, shipyards and chandlers started servicing yachts and other pleasure craft where once they built fishing schooners and other seagoing vessels. Captains' houses, fishing company warehouses, and corner stores were turned into swank bed and breakfasts, upscale restaurants, and tourist shops, all proclaimed as historical properties by plaques outside their doors. The fisheries museum along the waterfront preserved the history of the early fishery, while rusting hulls of groundfish trawlers bobbed up and down in the harbour adjacent to the quay where the first trawler – the *Cape North* – was moored.

Local events were transformed to reflect the changing economic realities. For example, the Lunenburg Fishermen's Reunion used to be called the Lunenburg Fishermen's Exhibition, featuring boats, fishing gear, and survival suits for people actively involved in the industry. Now, by gathering together families still engaged in the fishery for picnics with family, returning friends, and tourists, the event celebrates a way of life that is rapidly disappearing from the coast.

This romanticization of the past is not new. For example, the folklorist Helen Creighton portrayed the 'fishing folk' in the 1920s through 1960s as kind and simple people – men of steel who went out in wooden boats to do battle with the sea (McKay 1994). What was new was the extensive and systematic use of this image to sell the area through tourist brochures, community events, the print media, and productions for TV and film. Even *Good Morning, America*, broadcast from the wharves of Lunenburg proclaiming the town's advantages as a tourist site.

But the romantic portrayal of the past has little to do with the stark economic realities that have plagued the Lunenburg area. The message sent by the Federal government through their relief package, The Atlantic Groundfish Strategy, which came to an end in the summer of 1998, had been loud and clear: Get out of the fishery, retrain, develop alternative industries – most notably tourism – and resettle in areas of the country where there were jobs. But both 'cures' for these economic ills, emigration and alternative employment, had substantial side effects.

This economic restructuring also led to a social restructuring. The distribution of wealth within the community changed dramatically. Although there had been a reinvestment of 'old town' money derived from the fisheries and other related industries, there has also been an influx of 'new' money from outside the town. Entrepreneurs from Halifax, many of them emigrants from the U.S., joined with retirees and exurbanites from Central Canada, some of them Maritimers returning home to invest in Lunenburg's new economy. Investment by tourists and entrepreneurs resulted in increased values in houses, land, and rental property, but for many local long-term residents, any benefits from these increased property values could not be realized unless they sold their current homes and moved elsewhere. Financially hard-pressed fishing families were selling their houses and leaving, or else staying in the area but in rental accommodations or with family.

Other factors also led to an increased cost of living for long-term local residents. Tourists and entrepreneurs demanded particular facilities and speciality items not 'traditionally' available in the area. By creating these new demands for services and commodities, new entrepreneurial possibilities arose, but other commodities used predominately by long-term residents had been displaced and/or replaced by more expensive alternatives. People with disposable income welcomed this influx of more upscale consumer goods, but for some these new goods remained well beyond their economic reach. Inflation in the price of basic commodities and the replacement of these goods with the more expensive alternatives preferred by tourists meant that financial pressure on low-income families intensified. Moreover, in the winter many facilities closed, forcing locals to travel to shopping centres, notably to Halifax, to buy necessities.

Members of fishing-dependent households sought work in the tourist industry to supplement or replace the incomes they derived from fishing, but these new jobs, associated with the service sector, differed so fundamentally from their previous employment that such transitions were extremely difficult. Fishing-related jobs were relatively high paying, year-round employment for men on the boats, or full-time year-round or seasonal shift employment for men and women in the fish processing plants. As well, many of the people laid off from the fishery had low educational levels, and the jobs available to them in the tourist sector tended to be seasonal, part-time/split-shift work at minimum wage, and were usually more suitable for women. Moreover, during the winter months there were few employment opportunities available in the area. This de-skilling of the labour force also had special repercussions in the households of this male-oriented fishing culture. This transformation of employment opportunities, and the de-skilling of work have exacerbated the economic problems facing fishing families and other long-term residents.

All of these changes have created, in turn, tension between long-term residents and newcomers, conflicts that have been played out in many different ways within the communities. For example, some municipal by-laws were modified in favour of the new economic activities associated with tourism and to the disadvantage of other, more established activities associated with the fishing industry. In particular, conflicts have arisen between new residents and commercial fishing enterprises. Many tourists see small brightly coloured wooden fishing boats as quaint and romantic, but living beside a fully operational fishing enterprise brings strange smells and sounds. Tourist enterprises, bed and breakfast owners and motel operators successfully closed two fishing enterprises in one town by making changes in the town's by-laws. Such conflicts, and worse, are likely as tourism increases, especially if the fishery continues to decline.

The North Atlantic Fisheries Crisis was not unique; the fisheries are in decline around the world (McGoodwin 1990). Coastal communities ranging from Havana, Cuba, to Reykjavik, Iceland, in the Atlantic, and from Borocay, the Philippines, in the Pacific, to Split, Croatia, in the Mediterranean – all previously dependent on marine resources – have now turned to tourism, the largest and most expansive industry in the world, to cure their ailing economies. But in all these cases, this economic remedy has come at a price, including degradation of the environment, increased emigration, higher costs of living, more community conflict, the feminization and de-skilling of labour, the commodification of culture, and the invention of tradition. All of these costs bear on the Lunenburg situation.

The most recent studies of fishing-dependent communities of the North Atlantic rim, notably by Apostle and co-workers (1999), and Newell and Ommer (1998), focused on the restructuring of the fishing industry and the sustainability of these communities and their maritime environments. Few researchers, with the exception of anthropologists Anne Marie Powers (1998), Karen Szala-Meneok and Kara McIntosh (1996) in the case of Newfoundland, and historian Ian McKay (1994) and his graduate students in the case of Nova Scotia, looked at the social or cultural impacts of tourism on communities in Atlantic Canada. To date, no one has examined in-depth the role tourism has played in restructuring an Atlantic Canadian fishing-dependent community.

Whatever the future holds for the fishing-dependent households of Nova Scotia, the marine resource crisis of the 1990s had a profound effect on how these people make a living. Some left the industry; others modified their livelihood strategies to make a go of it. But most of us who visit the area only see the picturesque harbours, the romantic setting, and the museum depiction of fishing as simply a phase of the economic development of Atlantic Canada. We do not see the men and women who continue to struggle to support themselves and their households in an industry replete with physical dangers and economic risks.

Appendix

Table 1 Demographic Profiles of Deep-Sea Fishermen's Wives at Time of Interview

Interview number	Wife's name	Husband's name	Number of children	Employment outside the Home	Wife's age	Years married	Marital status
1	Amy	Adam	0	None	33	9	1st
2	Barbara	Bob	5	None	33	18	1st
3	Cathy	Cory	3	None	33	15	2nd
4	Donna	Dean	1	None	24	3	1st
5	Elizabeth	Eric	3	Nurse	51	29	1st
6	Frances	Fred	3	None	30	3	2nd
7	Gail	Gordon	2	Manager	35	7	1st
8	Holly	Harry	2	Nursing assistant	45	22	1st
9	Ina	Ivan	3	None	47	17	1st
10	Janet	Joe	5	Retired	52	30	1st
11	Karen	Ken	3	Retail sales	37	16	1st
12	Linda	Leon	2	None	30	9	1st
13	Margaret	Mark	1	Manager	36	15	1st
14	Nancy	Norman	3	Secretary	48	27	1st
15	Opal	Oscar	2	Waitress	41	24	1st
16	Paula	Peter	3	None	24	4	1st
17	Queenie	Quincy	2	Retired	49	29	1st
18	Ruth	Rob	3	Factory worker	38	20	1st
19	Sara	Stan	2	Owns business	38	18	Separated
20	Tara	Ted	2	Service sector	47	26	1st
21	Ursala	Ulysses	3	None	31	13	Separated
22	Vera	Vincent	0	Bank teller	26	4	Separated
23	Wendy	Walter	4	None	35	14	1st
24	Yvonne	Young	3	None	45	23	1st
25	Zareena	Zach	7	None	62	44	2nd

Table 2 Demographic Profiles of Coastal Fishermen's Wives at Time of Interview

Interview number	Wife's name	Husband's name	Number of children	Employment outside the Home	Wife's age	Years married	Marital status
26	Anne	Abraham	3	None	37	11	1st
27	Brenda	Bruce	2	None	48	18	1st
28	Claire	Colin	1	None	22	2	1st
29	Dorothy	Don	2	None	33	13	1st
30	Erica	Edward	4	None	42	30	1st
31	Faye	Frank	2	None	37	19	1st
32	Gilda	Graham	2	Factory work	38	19	1st
33	Hazel	Henry	1	Fisher	43	20	1st
34	Isabel	Ian	1	Salesperson	32	8	2nd
35	Jennifer	Jack	0	Service sector	24	7	1st
36	Kristen	Karl	2	Nurse	38	15	1st
37	Laurie	Len	1	None	32	7	1st
38	Martha	Michael	3	Service sector	33	12	1st
39	Norma	Nathan	3	Service sector	40	23	1st
40	Oriel	Obadiah	1	Owns business	39	18	1st
41	Pat	Paul	2	Teacher	46	23	1st
42	Quinsee	Quentin	2	Owns business	44	24	1st
43	Rosemary	Reed	3	Nurse's aide	57	32	1st
44	Susan	Stewart	4	Retail sales	38	7	2nd
45	Trudy	Tim	2	None	50	7	2nd
46	Unice	Uriah	3	None	58	36	1st
47	Virginia	Victor	1	None	33	10	1st
48	Wanda	William	4	Baby sits	61	39	1st
49	Yasmine	Yorick	2	None	43	2	2nd
50	Zoey	Zeb	3	Baby sits	33	15	1st

Notes

1. Introduction

1 I have arbitrarily assigned names to each woman interviewed and her family members. Throughout the book, quotations are identified consistently by these pseudonyms.
2 I use the term *fishermen/fisherman* unless I am talking about a female fisher, because I want to be gender-specific: in Nova Scotia at the time of the study all deep-sea fishers were men, and over 90 per cent of coastal fishers were men. Only one woman in this study owned and operated a fishing vessel, and held fishing licences in her name. All the other women I interviewed who fished with their husbands saw themselves as 'helpers,' not fishers. Second, it would be wrong to make this study appear to be gender-free. This is a study of fishing-dependent households with a focus on *fishermen's wives*, it is not a study of fishers' spouses. It is the gender-based relationship that is at the heart of the study.
3 An overview of discussions about the role of women in the fishery can be found in four sources: in the edited monograph by Carmelita McGrath, Barbara Neis, and Marilyn Porter (1995), *Their Lives and Times, Women in Newfoundland and Labrador: A Collage*; in special journal issues of *Anthropologica* (1996) – 'Women in the Fisheries,' edited by Karen Szala-Meneok, and of *Women's Studies International Forum* (2000) – *Women and the Fisheries Crisis*, edited by Dona Davis and Siri Gerrard; and a variety of articles in the journal *Samudra*. An additional source of information is the conference proceedings of the workshop 'Gender, Globalization and the Fisheries' held in St John's, Newfoundland, 6–12 May 2000.
4 I will not summarize the reasons for the crisis in the fishery, nor the government studies leading up to the crisis. For a summary and analysis of these questions see Alan Christopher Finlayson (1994).

5 The decline in the fishery has inevitably restructured the economies of the local communities, and the government has not offered any relief to the local retailers or other people indirectly affected by the fishery closures.
6 This is not unique in the North American or Canadian contexts; see Day and Brodsky (1998) and Elson (1992) for comparative examples.
7 See Cole 1991; Diaz 1999; Escallier and Maneschy 1996; Hall-Arber 1996; Harrison 1995; Jansen 1997; MacDonald 1994; Medard and Wilson 1996; Munk-Madsen 1998; Neis and Williams 1997; Overa 1993; Pratt 1996; Skaptadottir, 1996, 1998; Toril-Pettersen 1996; Vijayan and Nayak 1996; S.B. Williams 1988; and S. Williams 1996.
8 I have tried to make these descriptions as true to life as possible, but they are an amalgamation of a number of interviews.
9 Some researchers further separate the independent fishery into midshore and coastal fleets based on length of vessel and type of gear (e.g., Apostle and Barrett 1992). The coastal vessels were under thirty-five feet in length, while the midshore vessels were over thirty-five feet but less than sixty-five feet in length. Other researchers identified the coastal fishery with petty commodity production and the deep-sea fishery with industrial wage labour production (e.g., Brym and Sacouman 1979).
10 I have used the term *key informant* in the anthropological sense, meaning a person who is an important source of information or knowledge, although I recognize that the term *informant* has negative overtones for some readers and has been equated with *informer*.
11 According to the 1991 census data the median household income in 1991 was $34,669 for Nova Scotia and $39,047 for all of Canada (Canada, Statistics Canada 1993: 216).

2. Living the Dream

1 This federal program became 'employment' insurance (EI) by an act of Parliament in June 1996.
2 For comparative information on Newfoundland coastal fishing-dependent households on the Great Northern Peninsula, see Sinclair and Felt 1992.
3 I have excluded one coastal/deep-sea household (Len and Laurie) from this statistical analysis.

3. Two Separate Worlds

1 Since the early 1980s, starting with the Kirby Report (Canada, Task Force on Atlantic Canada 1983) the federal government has promoted the deep-sea industrialized fishery and multinational vertically integrated companies over the independent coastal fishery.

2 'Just in time' describes a system of inventory control and industrial production management based on the Japanese *kanban* system. Workers receive materials from suppliers 'just in time' for scheduled manufacturing to take place. Line workers generally signal that they require materials by means of a card or a computerized request system (cf., Microsoft Bookshelf '98).
3 One strategy used by fishing companies to downsize their fleet was to 'double crew' fishing vessels. In these cases each vessel had two full crews which alternated fishing trips. The men's work schedule was normally two weeks at sea followed by two weeks onshore.
4 In my 1986 study on deep-sea fishermen, most men reported taking off specific trips in the fall to go deer or moose hunting. In addition to contributing meat to their household's food supply, the men saw hunting as a social occasion where they could participate in manly activities with male relatives and friends.
5 This topic is discussed in more detail in chapter 4 under, 'Women's Employment.'
6 I am not assuming that a shared daily experience always leads to intimacy; rather, that there is more opportunity for it to develop with daily interaction.
7 Personal telephone communication with Leo Brander, Policy and Economics Branch, Nova Scotia Department of Fisheries and Oceans, 16 April 1996.
8 Regressive change meant substituting paid employment with other activities, either non-domestic or traditional male domestic chores (Wheelock 1990, 117).

4. Running the Household

1 On the division of domestic labour see Berk 1985; Berk and Shih 1980; Geerken and Gove 1983; Meissner et al. 1975; Michelson 1985; and Szala 1972. For studies using in-depth interview material, see Gavron 1966; Horrell 1994; Horrel *et al.* 1994; Kome 1982; Lopata 1971; Luxton 1980; Luxton and Corman 2001; Luxton and Rosenberg 1986; Oakley 1974; Rubin 1976, 1994; and Wheelock 1990.
2 I use the term *washing clothes*, not *doing the laundry*. As men and children do more domestic labour they often wash their own clothes, while women continue to have responsibility for the household laundry such as sheets, towels, tea towels, and other linens.
3 For a discussion of national patterns of divisions of household labour by sex, see Fredrick 1995 and Jackson 1992.
4 These findings are similar to those reported by Wheelock 1990 and Morris 1989, 1995.

5 Brayfield (1995) explored the relationship of men's work schedules to their ability to care for their children.
6 All of these studies, plus the Davis, Jentoft, and Thiessen study (1992) discussed in chapter 2, examined the division of labour in coastal households only.
7 Dona Davis's (1983b) research in a coastal fishing village in south-west Newfoundland (1977–80) reported that the previously strict division of labour based on gender was being eroded (26). Her study indicated that women's integration into full-time wage labour broke down the rigid gender-based division of labour. However, this fishing village has subsequently been dramatically affected by the cod moratorium: the fish plant has been closed and the boats tied up. The previous strict division of labour based on gender has returned (Davis, personal communication 1998).
8 Sinclair and Felt's (1992) research in isolated fishing outports on the Great Northern Peninsula of Newfoundland indicated that division of labour in these communities remained gender based. They argued that 'neither the presence of children in the home nor the employment of wives outside the home leads to significant changes in men's responsibility for domestic tasks' (56). Their research suggests that 'the survival of a fishing culture with its clear division of male and female tasks, the continuance of seasonal and part-time employment ... occupational pluralism requiring a diffuse set of gender specific skills (from motor repair and carpentry to sewing and "putting up" food), and a significant informal economy in which a wide range of goods and services are exchanged all serve to reinforce an historically grounded, gender society where peer groups continue to be a major point of reference in people's lives' (69). The livelihood strategies employed by households in these outport cultures require a set of skills which historically have been gender based. The isolation of these communities reinforces peer pressure to maintain these traditional forms of labour.
9 On men's resistance to engage in domestic labour, see Kome 1982; McMahon 1999; and Wheelock 1990. In *Negotiating Family Responsibilities* Finch and Mason (1993) explored the obligations and responsibilities associated with family life. Their study goes into great detail about how assistance within families was negotiated through time rather than given automatically. Men's resistance to engaging in domestic labour was part of this negotiating process. Gershuny and co-authors (1994) also argued this.

5. Family, Friends, Acquaintances

1 For an in-depth discussion of reciprocity and exchange within families, see Finch 1989.

2 These studies have not focused solely on the coastal sector – vessels less than forty-five feet – nor have researchers (e.g., Davis 1985; Apostle and Barrett 1992) differeniated the different sectors of the fishery – coastal, deep-sea and mid-shore (45 to 65 feet).
3 The language of employment is symptomatically ambigious here. 'Helpers' represent formal employment – paid work. Household members, such as wives or sons who 'help out,' but as unpaid labour, are not included in the discussion of 'who works for you.' Thus the unpaid participation of household members is excluded from these numbers.
4 Wellman 1992 discusses similar trends for other occupational groups of working class men.
5 It should be noted that bingo and lotteries not only promise solutions to financial problems; they also relieve the boredom of everyday life in a socially acceptable way.

6. Just Having Fun

1 These types of constraints are not unique to fishing-dependent households. For comparative material, see Chambers 1986; Firestone and Sheldon 1988; and Shaw 1985.
2 For comparative studies of home-based leisure, see Bella 1990, 1992; and Glyptis and Chambers 1982.
3 Various studies have looked at alcohol consumption in male blue-collar culture. See Hingson, Mangione, and Barrett 1981; Janes and Ames 1989; Le Masters 1975; and Trice and Sonnestuhl 1998.
4 I used the same definitions as those used in the National Alcohol and Other Drug Survey (NADS) conducted by Health and Welfare Canada in 1989, under the supervision of Marc Eliany. See Eliany et al. 1992 for particulars.
5 For results of the NADS survey and comparative material, see Eliany et al. 1992.
6 For a general discussion of the relationship between stress and high-risk drinking, see Cooper et al. 1990, and Neff and Baquer 1982. In the case of the fishery, Rix, Hunter, and College (1991) argued that drinking before a trip lessens some fishermen's anxiety about leaving their homes and families.

7. Going to Work

1 For a general discussion of this phenomenon see Brannen and Moss 1987.
2 Table 4.7 in chapter 4 shows that wives of deep-sea fishermen returned to the workforce sooner than their coastal counterparts.
3 I arbitrarily decided to present the data this way. The coastal sample results

had been consistently presented first in the other tables, so they are the baseline here.

4 For comparison with other studies of blue-collar households, see Barber 1992; Luxton and Corman 2001; and Wallman 1984.

8. Our Money, Your Money, My Money

1 For a more detailed discussion concerning the Nova Scotia Workers' Compensation Board and its policy, see Binkley 1995b, chap. 6.
2 Jan Pahl's typology of household allocation systems informs my understanding of household management systems (Pahl 1983: 245–9; 1989: 77–8). Lydia Morris (1989: 451) modified Pahl's typology to use them in a gender-free analysis. Carolyn Vogler (1994) also modified Pahl's typology to apply to no-earner and dual-earner households.

References

Allison, Charlene, Sue-Ellen Jacobs, and Mary A. Porter. 1989. *Winds of Change: Women in the Northwest Commercial Fishing.* Seattle: University of Washington Press.

Andersen, Raoul, ed. 1979. *North Atlantic Maritime Cultures.* The Hague: Mouton Press.

Andersen, Raoul, and Cato Wadel, eds. 1972. *North Atlantic Fishermen: Anthropological Essays on Modern Fishing.* St John's, Nfld.: ISER Books.

Antler, Ellen. 1977. 'Women's Work in Newfoundland Fishing Families.' *Atlantis* 2(2):106–13.

– 1982. 'Fishermen, Fisherwoman, Rural Proletariat: Capitalist Commodity Production in the Newfoundland Fishery.' PhD diss., University of Connecticut.

Apostle, Richard, and Jean Barrett, eds. 1992. *Emptying Their Nets: Small Capital and Rural Industrialization in the Nova Scotia Fishing Industry.* Toronto: University of Toronto Press.

Apostle, Richard, Jean Barrett, Petter Holm, Svein Jentoft, Leigh Mazany, Bonnie McCay, and Knut Mikalsen. 1998. *Community, State and Market on the North Atlantic Rim: Challenges to Modernity in the Fisheries.* Toronto: University of Toronto Press.

Apostle, Richard, Leonard Kasdan, and Art Hanson. 1985. 'Work Satisfaction and Community Attachment among Fishermen of Southwest Nova Scotia.' *Canadian Journal of Fisheries and Aquatic Sciences* 42:256–67.

Appadurai, Arjun. 1990. 'Disjuncture and Difference in the Global Economy.' *Theory, Culture and Society* 7:295–310.

Arnason, Ragnar, and Lawrence Felt. 1995. *The North Atlantic Fisheries: Successes, Failures, and Challenges.* Charlottetown, P.E.I.: Institute of Island Studies.

Barber, Pauline Gardiner. 1992. 'Household and Workplace Strategies in

"Northfield.'" In *Emptying Their Nets: Small Capital and Rural Industrialization in the Nova Scotia Fishing Industry*, edited by Richard Apostle and Gene Barrett. Toronto: University of Toronto Press, 272–99.

Bella, Leslie. 1990 'Women and Leisure: Beyond Androcentrism.' In *Understanding Leisure and Recreation: Mapping the Past, Charting the Future*, edited by E. Jackson. State College: Venture Press, 151–79.

– 1992. *The Christmas Imperatives: Leisure, Family and Women's Work*. Halifax: Fernwood Press.

Benjamin, Jessica. 1988. *The Bonds of Love: Psychoanalysis, Feminism and the Problem of Domination*. New York: Pantheon Press.

Berk, Richard. 1979. *Labor and Leisure at Home: Content and Organization of the Household Day*. Beverly Hills, Calif.: Sage.

Berk, Susan. 1985. *The Gender Factory: The Apportionment of Work in American Households*. New York: Plenum Press.

Berk, Sarah, and A. Shih. 1980. 'Contributions to Household Labour: Comparing Wives' and Husbands' Reports.' In *Women and Household Labour*, edited by Sarah Berk. Beverly Hills, Calif.: Sage, 191–227.

Binkley, Marian. 1994 *Voices From Off Shore: Narratives of Risk and Danger in the Nova Scotian Deep-Sea Fishery*. St John's, Nfld.: ISER Books.

– 1995a. 'Lost Moorings: Offshore Fishing Families Coping with the Fisheries Crisis.' *Dalhousie Law Journal* 18(1):84–95.

– 1995b. *Risks, Dangers, and Rewards*. Montreal: McGill-Queen's University Press.

– 1996. 'Nova Scotian Fishing Families Coping with the Fisheries Crisis.' *Anthropologica* 38(2):197–220.

Binkley, Marian, and Victor Thiessen. 1988. 'Ten Days a "Grass Widow" – Forty-Eight Hours a Wife: Sexual Division of Labour in Trawlermen's Households.' *Culture* 8(2):39–50.

Brannen, Julia, and Peter Moss. 1987. 'Dual Earner Households: Women's Financial Contributions after the Birth of the First Child.' In *Give and Take in Families: Studies in Resource Distribution*, edited by Julia Brannen and G. Wilson. London: Allen and Unwin, 75–95.

Brayfield, April. 1995. 'Juggling Jobs and Kids: The Impact of Employment Schedules on Fathers' Caring for Children.' *Journal of Marriage and the Family* 57 (May):321–32.

Brym, Robert, and James Sacouman, eds. 1979. *Underdevelopment and Social Movements in Atlantic Canada*. Toronto: University of Toronto Press.

Canada. Department of Fisheries and Oceans. 1991. *Canadian Fisheries Annual Statistical Review, 1986*. Vol. 19. Ottawa: Department of Fisheries and Oceans, Economic Analysis and Statistics Division, Economic and Commercial Analysis Directorate.

- Statistics Canada. 1993. *Census of Canada: Selected Income Statistics, 1991.* Ottawa: Industry, Science and Technology Canada.
- Statistics Canada. 2000. *Women in Canada: A Statistical Report, 2000.* Ottawa: Industry, Science and Technology Canada.
- Task Force on Atlantic Canada. 1983. *Navigating Troubled Waters: A New Policy for the Atlantic Fisheries* [The Kirby Report]. Ottawa: Supply and Services.

Canadian Institute for Research on Regional Development (CIRRD). 1995. *Economic Adjustment in Selected Coastal Communities* [The Savoy Report]. Fredericton, N.B.: CIRRD.

Chambers, Deborah. 1986. 'The Constraints of Work and Domestic Schedules on Women's Leisure.' *Leisure Studies* 5:309–25.

Christiansen-Ruffman, Linda. 1995. 'Researching Women's Organizations in the Labrador Straits: Retrospective Reflections.' In *Their Lives and Times, Women in Newfoundland and Labrador: A Collage*, edited by Carmelita McGrath, Barbara Neis, and Marilyn Porter. St John's, Nfld.: Killick Press, 249–63.

Cohen, Erik. 1988. 'Authenticity and Commoditization in Tourism.' *Annals of Tourism Research* 15:371–86.

Cole, Sally. 1991. *Women of the Praia: Work and Lives in a Portuguese Coastal Community.* Princeton: Princeton University Press.

Connelly, Patricia, and Martha MacDonald. 1985. 'Women and Development: The More Things Change, the More They Stay the Same.' *Women and Offshore Oil.* Conference Paper, No. 2, St John's, Nfld.: Institute of Social and Economic Research, 392–428.

Cooper, M. Lynne, Marcia Russell, and Michael R. Frone. 1990. 'Work Stress and Alcohol Effects: A Test of Stress-Induced Drinking.' *Journal of Health and Social Behavior* 31(3):260–76.

Copes, Parzival. 1986. 'A Critical Review of the Individual Quota as a Device in Fisheries Management.' *Land Economics* 62:278–91.

Danowski, Fran. 1980. *Fishermen's Wives: Coping with an Extraordinary Occupation.* Marine Bulletin 37. Kingston, R.I.: International Center for Marine Resource Development.

Davis, Anthony. 1985. '"You're Your Own Boss": An Economic Anthropology of Small Boat Fishing in Port Lameron Harbour, Southwestern Nova Scotia.' PhD diss., University of Toronto.

Davis, Dona Lee. 1983a. *Blood and Nerves: An Ethnographic Focus on Menopause.* St John's, Nfld.: ISER Books.
- 1983b. 'The Family and Social Change in the Newfoundland Outport.' *Culture* 3(1):19–31.
- 1995. 'Women in an Uncertain Age: Crisis and Change in a Newfoundand Community.' In *Their Lives and Times, Women in Newfoundland and Labrador:*

A Collage, edited by Carmelita McGrath, Barbara Neis, and Marilyn Porter. St John's, Nfld.: Killick Press, 279–95.

Davis, Dona Lee, and Siri Gerrard, eds. 2000. 'Special Issue: Women and the Fisheries Crisis.' *Women's Studies International Forum* 23(3):279–372.

Davis, Dona Lee, and Jane Nadel-Klein. 1992 'Gender, Culture, and the Sea: Contemporary Theoretical Approaches.' *Society and Natural Resources* 5: 135–47.

Day, Shelagh, and Gwen Brodsky. 1998. *Women and the Equality Deficit: The Impact of Restructuring Canada's Social Programs*. Ottawa: Status of Women.

Deem, Rosemary. 1986. *All Work and No Play? The Sociology of Women and Leisure*. Milton Keyes, U.K.: Open University Press.

Diaz, Estrella. 1999. 'Women Workers Feeling Insecure.' *Samudra* 22 (April): 37–9.

Donaldson, Mike. 1991. *Time of Our Lives: Labour and Love in the Working Class*. Sydney, Australia: University of Wollongong.

Eliany, Marc, Norman Giesbretch, Mike Nelson, Barry Wellman, and Scot Wortley. 1992. *Alcohol and Other Drug Use by Canadians: A National Alcohol and Other Drug Survey (1989)*. Technical Report. Ottawa: Queen's Printer.

Elson, Diane. 1992. 'From Survival Strategies to Transformation Strategies: Women's Needs and Structural Adjustment.' In *Unequal Burden: Economic Crises: Persistent Poverty and Women's Work*, edited by Shelly Feldman and Lourdes Beneria. Boulder, Colo.: Westview Press, 26–48.

Escallier, Christine, and Maria Cristina Maneschy. 1996. 'Women Net Weaver: Weaving a Living.' *Samudra* 14 (March):2–4.

Faris, James C. 1966. *Cat Harbour: A Newfoundland Fishing Settlement*. St John's, Nfld.: ISER Books.

Finch, Janet. 1989. *Family Obligations and Social Change*. Cambridge, U.K.: Polity Press.

Finch, Janet, and Jennifer Mason. 1993. *Negotiating Family Responsibilities*. London: Routledge.

Finlayson, Alan Christopher. 1994. *Fishing for Truth: A Sociological Analysis of Northern Cod Stock Assessment from 1977 to 1990*. St John's, Nfld.: ISER Books.

Firestone, Juanita, and Beth A. Sheldon. 1988. 'An Estimation of the Effects of Women's Work on Available Leisure Time.' *Journal of Marriage and the Family* 9(4):478–95.

Fox, Bonnie. 1997. 'Reproducing Difference: Changes in the Lives of Partners Becoming Parents.' In *Feminism and Families: Critical Policies and Changing Practices*, edited by Meg Luxton. Halifax: Fernwood Press, 142–61.

Frederick, Judith. 1995. *As Time Goes By ... Time Use of Canadians*. Ottawa: Statistics Canada, Ministry of Industry.

Gavron, Hannah. 1966. *The Captive Wife: Conflicts of Household Mothers.* London: Routledge and Kegan Paul.

Geerken, M., and W.R. Gove. 1983. *At Home and at Work: The Family Allocation of Labor.* Beverly Hills: Sage.

Gerrard, Siri. 1995. 'When Women Take the Lead: Changing Conditions for Women's Activities, Roles and Knowledge in North Norwegian Fishing Communities.' *Social Science Information* 34(4):593–631.

Gershuny, Jonathan, Michael Godwin, and Sally Jones. 1994. 'The Domestic Labour Revolution: A Process of Lagged Adaptation.' In *The Social and Political Economy of the Household,* edited by Michael Anderson, Frank Bechhofer, and Jonathan Gershuny. Oxford: Oxford University Press, 151–97.

Gerstel, Naomi, and Harriet Gross. 1984. *Commuter Marriage: A Study of Work and Family.* New York: Guilford Press.

Gloor, Daniela. 1992. 'Women versus Men? – The Hidden Differences in Leisure Activities.' *Society and Leisure* 15(1):39–60.

Glyptis, Susan. 1989. *Leisure and Unemployment.* Milton Keyes, U.K.: Open University Press.

Glyptis, Susan, and Deborah Chambers. 1982. 'No Place Like Home.' *Leisure Studies* 1:147–62.

Hall-Arber, Madeleine. 1996. 'Hear Me Speak: Italian and Portuguese Women Facing Fisheries Management.' *Anthropologica* 38:221–48.

Harrington, Maureen, Don Dawson, and Pat Bella. 1992. 'Objective Constraints on Women's Enjoyment of Leisure.' *Society and Leisure* 15(1):203–21.

Harrison, Elizabeth. 1995. 'Fish and Feminists.' *IDS Bulletin* 26(3):39–47.

Harvey, David. 1989. *The Condition of Postmodernity.* Oxford: Basil Blackwell Press.

Hingson, Ralph, Thomas Mangione, and Jane Barrett. 1981. 'Job Characteristics and Drinking Practices in the Boston Metropolitan Area.' *Journal of Studies on Alcohol* 42(9): 725–38.

Horbulewicz, Jan. 1972. 'The Parameters of the Psychological Autonomy of Industrial Trawler Crews.' In *Seafarer and Community,* edited by Peter Fricke. London: Croom Helm, 67–84.

Horrell, Sara. 1994. 'Household Time Allocation and Women's Labour Force Participation.' In *The Social and Political Economy of the Household,* edited by Michael Anderson, Frank Bechhofer, and Jonathan Gershuny. Oxford: Oxford University Press, 198–224.

Horrell, Sara, Jill Rubery, and Bendan Burchell. 1994. 'Working-Time Patterns, Constraints and Preferences.' In *The Social and Political Economy of the Household,* edited by Michael Anderson, Frank Bechhofer, and Jonathan Gershuny. Oxford: Oxford University Press, 100–32.

Ilcan, Suzan. 1986. 'Women and Casual Work in the Nova Scotian Fish Processing Industry.' *Atlantis* 11(2):23–34.

Jackson, Chris. 1992. 'The Value of Housework in Canada.' In *National Income and Expenditure Accounts*. Ottawa: Statistics Canada, Ministry of Industry, xxxiii–lii.

Janes, Craig R., and Genevieve Ames. 1989. 'Men, Blue Collar Work and Drinking: Alcohol Use in an Industrial Subculture.' *Culture, Medicine and Psychiatry* 13(3):245–74.

Jansen, Erik. 1997. *Rich Fisheries–Poor Fisheries: Some Preliminary Observations About the Effects of Trade and Aid in the Lake Victoria Fisheries*. Report No. 1. IUCN Eastern Africa Programme, Socio-economics of the Lake Victoria Fisheries. Nairobi: IUCN, The World Conservation Union.

Kelly, Ken. 1993. 'Atlantic Canada Reels under Fishery Closures.' *National Fisherman*. (December):14–15.

Kome, Penny. 1982. *Somebody Has Got to Do It: Whose Work Is Housework?* Toronto: McClelland & Stewart.

Le Masters, E.E. 1975. *Blue-Collar Aristocrats: Life-Styles at a Working-Class Tavern*. Madison: University of Wisconsin.

Lopata, Helena. 1971. *Occupation: Housewife*. London: Oxford University Press.

Luxton, Meg. 1980. *More Than a Labour of Love*. Toronto: Women's Press.

Luxton, Meg, and June Corman. 2001. *Getting By in Hard Times: Gendered Labour at Home and on the Job*. Toronto: University of Toronto Press.

Luxton, Meg, and Harriet Rosenberg. 1986. *Through the Kitchen Window: The Politics of Home and Family*. Toronto: Garamond Press.

MacCannel, Dean. 1989. *The Tourist: A New Theory of the Leisure Class*. New York: Schocken Books.

MacDonald, Martha. 1994. 'Restructuring in the Fishing Industry in Atlantic Canada.' In *The Strategic Silence: Gender and Economic Policy*, edited by Isabella Bakker. London: Zed Books, 91–102.

McCay, Bonnie Jean. 1995. 'User Participation in Fisheries Management: Lessons Drawn from International Experiences.' *Marine Policy* 19(3):227–46.

McCay, Bonnie Jean, Carolyn Creed, Alan Christopher Finlayson, Richard Apostle, and Knut Mikalsen. 1995. 'Individual Transferable Quotas (ITQs) in Canadian and American Fisheries.' *Ocean and Coastal Management* 28(1–3):85–116.

McCrone, David. 1994. 'Getting By and Making Out in Kirkcaldy.' In *The Social and Political Economy of the Household*, edited by Michael Anderson, Frank Bechhofer, and Jonathan Gershuny. Oxford: Oxford University Press, 68–99.

McGoodwin, James R. 1990. *Crisis in the World's Fisheries: People, Problems and Policies*. Stanford: Stanford University Press.

McGrath, Carmelita, Barbara Neis, and Marilyn Porter, eds. 1995. *Their Lives and Times, Women in Newfoundland and Labrador: A Collage.* St John's, Nfld.: Killick Press.

McKay, Ian. 1994. *The Quest of the Folk: Antimodernism and Cultural Selection in Twentieth-Century Nova Scotia.* Montreal: McGill-Queen's University Press.

McMahon, Anthony. 1999. *Taking Care of Men: Sexual Politics in the Public Mind.* Cambridge: Cambridge University Press.

Medard, Modesta, and D.C. Wilson. 1996. 'Changing Economic Problems for Women in the Nile Perch Fishing Communities on Lake Victoria.' *Anthropologica* 38(2):149–72.

Meissner, Martin, Elizabeth Humphries, Scot Meis, and William Scheu. 1975. 'No Exit for Wives: Sexual Division of Labour and the Culmination of Household Demands.' *Canadian Review of Sociology and Anthropology* 12(4):424–39.

Michelson, William. 1985. *From Sun to Sun: Daily Obligations and Community Structure in the Lives of Employed Women and Their Families.* Ottawa: Rowman and Allenheld.

Microsoft. 1998. *Microsoft Bookshelf '98.* Redwood, Wash.: Microsoft.

Morris, Lydia. 1989. 'Household Strategies: The Individual, the Collectivity and the Labour Market – the Case of Married Couples.' *Work, Employment and Society* 3:447–64.

– 1990. *The Workings of the Household.* Cambridge, U.K.: Polity Press.

– 1995. *Social Divisions: Economic Decline and Social Structural Change.* Cambridge, U.K.: University of Cambridge Press.

Munk-Madsen, Eva. 1998. 'The Norwegian Fishing Quota – Another Patriarchal Construction?' *Society and Natural Resources* 21:229–40.

– 2000. 'Wife the Deckhand, Husband the Skipper: Authority and Dignity among Fishing Couples.' *Women's Studies International Forum* 23(3):333–42.

Nadel-Klein, Jane, and Dona Lee Davis, eds. 1988. *To Work and to Weep: Women in Fishing Economies.* St John's, Nfld.: ISER Books.

Nash, June. 1994. 'Global Integration and Subsistence Insecurity.' *American Anthropologist* 96(1):7–30.

Neff, James Alan, and Baquer A. Husaini. 1982. 'Life Events, Drinking Patterns and Depressive Symptomatology.' *Journal of Studies of Alcohol* 43:301–18.

Neis, Barbara. 1996. 'Cut Adrift.' *Samudra Report* 16 (November):35–9.

Neis, Barbara, Lawrence Felt, D.C. Schnider, R. Haerdrich, Jeff Hutchings, and J. Fischer. 1996. *Northern Cod Stock Assessment: What Can Be Learned from Interviewing Resource Users?* Dartmouth, N.S.: Northwest Atlantic Fisheries Organization.

Neis, Barbara, and Susan Williams. 1997. 'The New Right, Gender and the Fisheries Crisis: Local and Global Dimensions.' *Atlantis* 21(2):47–63.

Newell, Diane, and Rosemary E. Ommer, eds. 1999. *Fishing Places, Fishing People: Traditions and Issues in Canadian Small-Scale Fisheries.* Toronto: University of Toronto Press.

Northwest Atlantic Fisheries Organization. 1989. *Scientific Council Summary Document.* No. N1699. Dartmouth, N.S.: NAFO.

Oakley, Anne. 1974. *The Sociology of Housework.* London: Martin Robinson.

Overa, Ragnhild. 1993. 'Wives and Traders: Women's Careers in Ghanaian Canoe Fisheries.' *Maritime Anthropological Studies* 6(1, 2):110–35.

Pahl, Jan. 1983. 'The Allocation of Money and the Restructuring of Inequality within Marriage.' *Sociological Review* 31(2):237–62.

– 1989. *Money and Marriage.* New York: St Martin's Press.

Pearlin, Leonard, and Carmi Schooler. 1978. 'The Structure of Coping.' *Journal of Health and Social Behavior* (19):2–21.

Poggie, John J. Jr. 1980. 'Ritual Adaptation to Risk and Technological Development in Ocean Fisheries: Extrapolations from New England.' *Anthropological Quarterly* 53(1):122–9.

Porter, Marilyn. 1983. 'Women and Old Boats: The Sexual Division of Labour in a Newfoundland Outport.' In *Public and Private: Gender and Society*, edited by E. Garminkow et al. London: Heinemann and BSA, 1–77.

– 1985a. '"A Tangly Bunch": The Political Culture of Outport Women in Newfoundland.' *Newfoundland Studies* 1:77–90.

– 1985b. '"She Was Skipper of the Shore-Crew": Notes on the History of the Sexual Division of Labour in Newfoundland.' *Labour/Le Travail* 15:105–24.

Powers, Anne Marie. 1998. 'The Construction of Cultural Identity: Cabot 500 Year Celebrations and Come Home Celebrations in Newfoundland.' Paper presented at the British Association of Canadian Studies, Stoke-on-Trent, England.

Pratt, Marion. 1996. 'Useful Disasters: The Complexity of Response to Stress in a Tropical Lake Eco-System.' *Anthropologica* 38(2):125–48.

Rix, K. J.B., D. Hunter, and P.C. Colley. 1982. 'Incidence of Treated Alcoholism in North-East Scotland, Orkney and Shetland Fishermen, 1966–1970.' *British Journal of Industrial Medicine* (39):11–17.

Rubin, Lillian. 1976. *Worlds of Pain: Life in the Working Class Family.* New York: Basic Books.

– 1994. *Families on the Fault Line: America's Working Class Speaks about the Family, the Economy, Race, and Ethnicity.* New York: Harper Collins.

Shaw, Susan. 1985. 'The Meaning of Leisure in Everyday Life.' *Leisure Sciences* 7(1):1–24.

Shaw, Susan, Arend Bowen, and John McCabe. 1991. 'Do More Constraints Mean Less Leisure? Examining the Relationship between Constraints and Participation.' *Journal of Leisure Research* 23(4):286–300.

Shrybman, Steven. 1999. *A Citizen Guide to the World Trade Organization.* Ottawa: Canadian Centre for Policy Alternatives.

Sinclair, Peter R., and Lawrence F. Felt. 1992. 'Separate Worlds: Gender and Domestic Labour in an Isolated Fishing Region.' *The Canadian Review of Sociology and Anthropology* 29(1):55–71.

Skaptadottir, Unner Dis. 1996. 'Gender Construction and Diversity in Icelandic Fishing Communities.' *Anthropologica* 38(2):271–87.

– 1998. 'Coping with Marginalization: Localization in an Icelandic Fishing Community.' In *Coping Strategies in the North: Approaching Local Practices of Integration,* edited by Nils Aarseether and Jorgen Ole Baerenholdt. Copenhagen: Nordic Council of Ministers, 91–106.

Stiles, Geoffrey. 1971. 'Labour Recruitment and the Family Crew in Newfoundland.' In *North Atlantic Maritime Cultures: Anthropological Essays on Changing Adaptations,* edited by R. Andersen. The Hague: Mouton, 189–208.

Stonich, Susan C., J.R. Bort, and L.L. Ovares. 1997. 'Globalization of Shrimp Mariculture: The Impact of Social Justice and Environmental Quality in Central America.' *Society and Natural Resources* 10:161–79.

Szala, A., ed. 1972. *The Use of Time: Daily Activities of Urban and Suburban Populations in Twelve Countries.* The Hague: Mouton.

Szala-Meneok, Karen, ed. 1996. 'Women in the Fisheries.' *Anthropologica* 38(2):119–288.

Szala-Meneok, Karen, and Kara McIntosh. 1996. 'Craft Development and Development Through Crafts: Adaptive Strategies of Labrador Women in a Changing Fishery,' *Anthropologica* 38(2):249–70.

Thiessen, Victor, Anthony Davis, and Svein Jentoft. 1992. 'The Veiled Crew: An Exploratory Study of Wives' Reported and Desired Contributions to Coastal Fisheries Enterprises in North Norway and Nova Scotia.' *Human Organization* 51(4):342–52.

Thompson, Paul. 1985. 'Women in the Fishing: The Roots of Power between the Sexes.' *Comparative Study of Society and History* 27(1):3–32.

Toril-Pettersen, Liv. 1996. 'Crisis Management and Household Strategies in Lofoten: A Question of Sustainable Development.' *Sociologia Ruralis* 36(2):236–48.

Trice, Harrison M., and William J. Sonnestuhl. 1988. 'Drinking Behavior and Risk Factors Related to the Work Place: Implications for Research and Prevention.' *Journal of Applied Behavioral Science* 24(4):327–46.

United Nations Food and Agricultural Organization (UN-FAO). 1987. *World Nominal Catches.* Vol. 64. Rome: FOA.

– 1998. *Status of Marine Fisheries.* Rome: FOA.

Urry, John. 1990. 'Cultural Changes and Restructuring Tourism.' *The Tourist Gaze: Leisure and Travel in Contemporary Societies.* London: Sage, 82–103.

Vijayan, Aleyamma, and Nalini Nayak. 1996. *Women First: Report of the Women in Fisheries Programme of ICSF in India.* Samudra Dossier Women in Fisheries Series No. 2. Chennai, India: International Collective in Support of Fishworkers.

Vogler, Carolyn. 1994. 'Money in the Household.' In *The Social and Political Economy of the Household,* edited by Michael Anderson, Frank Bechhofer, and Jonathan Gershuny. Oxford: Oxford University Press, 225–66.

Wallman, Sandra. 1984. *Eight London Households.* London: Tavistock Publications.

Watts, Michael. 1992. 'Living under Contract: Work, Production, Politics and the Manufacturing of Discontent in a Peasant Society.' In *Reworking Modernity: Capitalism and Symbolic Discontents,* edited by Allen Pred and Michael Watts. New Brunswick, N.J.: Rutgers University Press, 65–105.

Wellman, Barry. 1985. 'Domestic Work, Paid Work and Net Work.' In *Understanding Personal Relationships: An Interdisciplinary Approach,* edited by S. Duck and D. Perlman. Beverly Hills, Calif.: Sage, 159–61.

– 1992. 'Men in Networks: Private Communities, Domestic Friendships.' In *Men's Friendships,* edited by Peter Norch. Newberry Park, Calif.: Sage, 74–114.

– 1999. 'From Little Boxes to Loosely Bounded Networks: The Privatization and Domestication of Community.' In *Sociology For the 21st Century: Continuities and Cutting Edges,* edited by Janet Abu-Lughod. Chicago: University of Chicago Press, 94–114.

Wheelock, Jane. 1990. *Husbands at Home: The Domestic Economy in a Post-Industrial Society.* London: Routledge.

Wilks, Alex. 1995. 'Prawns, Profit and Protein: Aquaculture and Food Production.' *The Ecologist* 25(2–3):120–5.

Williams, Stella B. 1988. 'Women's Participation in the Fishing Industry in Nigeria: A Review.' *African Notes* 3:51–4.

Williams, Susan. 1996. *Our Lives Are at Stake: Women and the Fishery Crisis in Newfoundland and Labrador.* ISER Report No. 11. St John's, Nfld.: ISER Books.

Index

Abuse, physical, 104, 133
Alcoholics Anonymous /Al-Anon, 128–9
Allowances: housekeeping, 178–81, 183; personal, 82, 178–80, 183
Atlantic Groundfish Strategy, The (TAGS): description of, 9; downsizing, effects of, on, 58; as income, 161; package taken, 161, 176; termination of, 9, 192; women, exclusion of, from, 10

Babysitting, 39, 76, 78, 92–4, 96–7, 101, 103, 120, 142, 152
Business, family, 14, 28, 42

Caregiver/caregiving, 43, 91, 133
Child care, 72–80; as constraint on employment, 73–5, 140–8, 152; costs of, 116; employment in, effects of downsizing on, 10; in formal sector, 75; in life course, 47, 72, 77, 91, 102; husband's involvement in, 39, 54–5, 58, 60–1, 78–80, 82; providers of, 75–7, 107; responsibilities shared with friends, 75–7, 101; as service, exchange of, 83, 93, 97; and wife's involvement in fishing enterprise, 76; as wife's primary responsibility, 22, 25, 28, 35, 38, 65, 74, 75, 77, 81, 86, 93; withholding of, as social control, 76, 103
Children: activities of, 110, 112, 113, 115–6, 120; demographics of, 17, 18, 20, 21, 72–3, 74; domestic labour, involved in raising of, 72, 91; father's absence, effects of, on, 51–2, 55; father's drinking, effects of, on, 127, 130, 131; father's involvement with, 39, 54–5, 58, 60–1, 78–80, 82, 113–14; financial, imperatives of, 166–7, 169, 176, 180, 182; and husband's needs, balanced with, 55, 112, 141–2, 150; mother's employment, as limitations on, 73–5, 134, 140–8, 152; support of, 91, 94, 96, 97; wife's primary responsibility, 22, 25, 28, 35, 38, 65, 71, 74, 75, 77, 93
Community: belonging, sense of, 54, 60–1, 62, 95, 102, 105–6, 122, 189; choice of, 106–8, 129; fishermen, laid off, increased involvement in,

216 Index

59, 60; fisheries crisis, effects of, on, 110–11; goods and services, exchange of, in, 116; leisure activities, in, 112, 114–17, 192; social control in, 103, 108–10; status within, 52; support network in, 92–3, 96, 130; values of, 11, 111, 146, 192–3

Community services: availability of, 104, 106, 137, 138, 146; post office, 69, 71; reduction of, 9, 11, 92; use of, 69, 70, 107, 115

Commodification, 191, 194

Daycare, 75–7, 92, 135, 138, 140–6

Department of Fisheries (provincial), 44

Department of Fisheries and Oceans (federal), 44

Division of labour, 66, 68, 93; in coastal fishing households, 22–3, 66, 67, 81, 84, 190; comparison with other studies, 80; in deep-sea fishing households, 59, 66–8, 81, 187, 190; reliance on other help, 71, 77

Domestic labour: chores, 66; —, traditional female, 62, 66, 71, 80–6; —, traditional male, 68, 69; —, feminization of, 71; family life cycle, relationship to, 65, 72, 91; fishery, type of, relationship to, 65, 69; —, coastal, 22, 23, 25, 66–7, 68; —, deep-sea, 52, 66–7; husband's employment status, relationship to, 57, 58, 59, 62, 65, 88; livelihood, relationship to, 65; reliance on others, for, 76, 86; women's employment, relationship to, 65, 73–80; —, children and childcare,

effects of, on, 73–6; and women's involvement in fisheries enterprise, 42, 65

Drinking: consumption, levels of, 121–3, 125; drunk drivers, 127, 131; and leisure activities, 56, 117, 121–3, 127, 129, 133, 189; negative aspects of, 56, 123–33, social, 119, 122–3, 189; taming process, 56–7, 102, 120, 125

Employment Insurance: direct payment of, 166; as income, 161, 162; reliance on, 9, 11, 26, 27, 92

Exchange, production for, 15

Exchange of goods and services, informal: within communities, 116; definition of, 91–3, 188; within families, 83, 87–8, 93–4, 96; between individuals, 87; of labour, 76; within support networks, 92–3, 101–2. *See also* 'Helping out'

Expenses: of coastal fishing enterprise, 32, 36–7, 39–40, 138, 146, 158; of household, 13, 165–6, 176, 179, 181–4; life cycle, relationship to, 147; personal, 183

Financial resources, household: access to, 82, 156, 165, 167–71, 173, 178, 180–2, 185; control of, 171–80; fisheries crisis, impact of, on, 166–7, 189–90; independence and autonomy with, 58, 150–1, 153–6, 177–80; management strategies for, 178–85; planning imperatives for, 163–7; pooled, 155, 168–72, 174, 178, 181, 183–4; and power relations, 72, 81–2, 84, 92, 151, 154, 178, 182–3, 188–9; segregated,

162–72, 174, 178; withholding as social control, 182
Fisheries, coastal: adaptations to, 16, 28; description of, 16, 24–8; fisheries crisis, response to, 25–6, 36–42; recruitment to, 23–4; share system in, 158; women's involvement in, 26–7, 29–36, 38–41
Fisheries, deep-sea: adaptations to, 16; description of, 16, 43, 44–7; fisheries crisis, response to, 47, 57–64; share system, 158; recruitment to, 44
Fisheries crisis, Nova Scotia: communities, effects of, on, 110–11; description of, 4, 6, 190; as economic restructuring, 5, 6, 11; fishers, effects of, on, 9; household financial resources, effects of, on, 166–7, 189–90; response to, 3, 5; —, in coastal fishing, 25–6, 36–42; —, in deep-sea fishing, 47, 57–64; —, governmental, 9–11; —, leaving the fishery, 40–2; women's paid employment, effects of, on, 6, 10, 58–9, 62–4, 84, 86, 156, 167, 188. *See also* Atlantic Groundfish Strategy, The
Fisheries crisis, world: description of, 5; women and their concerns, exclusion of, in, 10
Fishing-dependent households: coastal, 14, 22–3, 42, 66–7, 81, 190; comparison of, 14–15; deep-sea, 14–15, 43–4, 47–59, 64, 66–8, 81, 102, 120, 125, 187–8, 190
Fishermen, coastal: working conditions of, 20, 22–43
Fishermen, deep-sea: drinking by, 56–7, 102, 120, 125; on double-crewed vessels, 57–60, 64, 86; employment status, 9, 47, 57–64, 88–9, 147, 156, 188, 193; shore life, coping with, by, 54–5; working conditions of, 20, 46–7, 59
Friends/friendships: as couples, 68, 102, 103, 117–20; dynamic nature of, 101–3; men's, 48, 56–7, 61, 106, 125–6; social network, as part of, 91–3, 120–1; women's, 48, 52, 75–7, 96, 101–5, 107, 111–13, 117, 128–30, 150–1, 188–9

'Getting by,' 163–4, 189
Gender relations: understanding, of, 4; women, marginalization of, 1, 4, 10–11
Gill-netting, 27–8
Globalization: fishery, effects of, on, 4, 5, 11
Going on a tear. *See* Drinking
Gossip, 71, 103, 107–9
Government: cutbacks by, effects on women's work of, 11; fisheries crisis, response to, by, 9–11. *See also* Atlantic Groundfish Strategy, The

Halifax, 4, 6, 15, 57, 101, 104, 107, 110, 114, 118, 143, 152, 192
'Helping out': nature of, 65, 86, 87–9, 94, 188; in coastal fishery, 28–36; 39; households, 62, 71, 80–6, 95, 101

Income: declining, effects of, on, 92, 100, 167, 174; of double-crewed vessels, 86, 160; Employment Insurance, part of, 161, 162; fishermen's, unemployed, 9, 160; households, 19, 21, 59, 62, 81–2, 116,

122, 161–3, 178, 192; men's, 20–1, 159–60; —, in coastal fishing households, 20, 36, 159–60; —, in deep-sea fishing households, 20–1, 58, 159–61; TAGS (The Atlantic Groundfish Strategy), part of, 161; women's, 134–5, 137–8, 162–3; —, in coastal fishing households, 2, 32–5, 36, 38–9, 42, 85, 136; in deep-sea fishing households, 59, 64, 82, 84, 86, 136–7; —, dependency on, 153, 156, 177, 180; —, household, crucial to maintaining, 4, 138, 142, 147, 152, 156, 167, 183–5, 188, 193

Kin: gender segregation of, 93; locality of, 99–100, 107; matrilineal, 99; patrilineal, 99–100; reliance on, 23, 36, 72, 77, 93, 99–100

Labour: double day of, 84, 189; triple day of, 35, 84, 112, 189
Leisure activities: 18, 112–21; and alcohol, 56, 117, 121–3, 127, 129, 133, 189; and child care, need for, 76–7, 93, 140; with children, 78, 101–2, 110, 112, 113–16, 120, 140; choice of, factors affecting, 113–15, 182; community, availability of, in, 112, 114–17, 192; as couple, 68, 112–13, 117–20; men's, 54, 56, 57, 60, 61, 120; women's, 101–10, 112–17, 120, 141, 150–1, 189
Life course/life cycle: child care, relationship of, to, 47, 72, 77, 91, 102; domestic labour, relationship of, to, 65, 72, 91; expenses, changes of, during, 147; family responsibilities, changes of, during, 91, 96, 187–8; women: friendships of, during, 101–2, 188–9; —, in coastal fishery, involvement of, during, 26, 28, 35, 187; —, in deep-sea households, relationship to work, during, 47, 48; —, paid employment of, during, 134, 137, 141–8, 189
Livelihood: definition of, 16; income, effects of, on, 21; domestic labour, relationship to, 65
Lobstering, 26–7, 37–8, 40, 139
Longlining, 27–8
Lunenburg, 15, 17, 106–7, 114, 123, 190–4
Lunenburg Fishermen's Reunion, 191
Lunenburg Fishermen's Exhibition, 140, 191

'Making out,' economically, 163–4
Marine resources: cyclical nature of, 4, 37
Markets: for coastal fishery, 38; for deep-sea fishery, 45

'Necessary visiting,' 117–19
Neighbours, 71, 87; as support, 76, 87, 91–2, 101, 104, 105–10

Patrilocality, 100
Post office, 69, 71
Power relations, 72, 81–2, 84, 92, 151, 154, 178, 182–3, 188–9

Reciprocity, 188
Restructuring, 5, 6, 11, 58–9
Retraining, 9, 11, 152

Share system, 158, 185
Social benefits: access to, 4, 10; reduction of, 6, 9

Social control, 76, 93–4, 97, 101, 103–4, 108, 182, 189
Sports, 61, 104, 110, 112, 115
Subordination, of women to men, 49, 50, 82

TAGS. *See* Atlantic Groundfish Strategy, The
Tourists/tourism, 5, 10, 15, 24, 111, 190–4

Unemployment, 5, 6, 9, 58, 59
Unemployment Insurance. *See* Employment Insurance

Volunteer work, 91, 102, 104–5, 110, 114–16, 120, 140

Wage labour, 4, 6, 14, 22, 25, 36, 39, 84
Women's paid employment: childcare limitations to, 48, 73–5, 134, 140–8, 152; as crucial to maintaining household, 11, 34–6, 39, 133, 151–2; demographics of, 134–7; domestic labour, relationship of, to, 65, 73–80, 82, 140, 143–8, 189; fisheries crisis, effects of, on, 6, 10, 58–9, 62–4, 84, 86, 156, 167, 188; and friendship network, enhancement of, 103, 105, 120, 150; and husband's work schedule, modification of, 14, 43, 47, 112, 138–40, 141; and independence and autonomy, 58, 150–1, 153–6, 177–80; and labour, double day of, 84, 189; and labour, triple day of, 35, 84, 112, 189; leisure activities, relationship of, to, 121, 150–1; life course, relationship of, to, 141–8, 189; reasons for, 142, 146–57; and tensions in household, 53, 73, 145–6; via volunteer labour, 105; women's involvement in coastal fisheries, effects of, on, 28, 33–5, 36, 38, 138–9, 190
Women's unpaid labour: centrality to fisheries, 11, 188–9; —, coastal, 22, 28, 42; —, deep-sea, 43–4; effects of, on fisheries, 33–5, 36, 38; in fisheries enterprise and domestic labour, 42, 65; husband's work, defined by, 15